How To Buy An Affordable Electric Car

A Tightwad's Guide To EV Ownership

Matt DeLorenzo

Cover design by MiblArt

Nissan LEAF photography by Guy Spangenberg

Cartoons by Henry Payne

DEDICATION

To Charlie DeLorenzo (1952-2022): Big brother, best friend and partner in crime forever.

CONTENTS

FOREWORD

Whether you're looking to save the planet or just a couple of bucks, this book is for you. Electric cars are here. There's plenty of hype encouraging people to jump onto the EV bandwagon. Trouble is, if you want to get into an electric, the bulk of the choices seem to be high-dollar EVs from Tesla, GMC Hummer and startups like Rivian. At first blush, it appears the average car buyer is not invited to the EV party.

Rest assured, there are more affordable alternatives in the marketplace. But finding out about these attainable EVs from mainstream makers is no simple task. There's much to learn about this new technology before trading in your old gas burner for a shiny new electric one.

Electric vehicles are different. The devil is in the details when it comes to how far you can go, how long it takes to recharge, whether you can find a place to plug it in and, most important of all, what it costs to buy or own an electric car.

So, what's with the Tightwad angle? The guidance contained in these pages will show you ways to save while giving you the information you need to decide if an electric car is right for you.

Having spent many years writing about cars for publications like *Autoweek* and *Road & Track* and driving everything from Ferraris to Fords, I recently came to the realization that when it comes to my personal cars, I'm a Tightwad. If you look in my garage, you won't find an expensive collectible Alfa Romeo or a new Corvette. Instead, it's been a steady progression of affordable cars including an AMC Gremlin, Dodge Omni, Saturn SL, Honda Accord, Mazda

626 and a couple of PT Cruisers.

Two of our family's acquisitions include a 2008 Toyota Prius Hybrid and a 2018 Hyundai Ioniq Plug-in Hybrid. Both cars represent baby steps taken by the auto industry towards full electrification. I decided I should follow along by buying a 2022 Nissan LEAF S electric. In true Tightwad fashion, the LEAF is the lowest-priced electric car on the market. The experience of buying and driving that car informs a lot of the material you'll find here, along with tips on buying new or used cars gleaned from my eight years writing about the new car market for Kelley Blue Book.

There's a big debate on the current and future role of electric vehicles in the transportation fleet. Some believe that we should get off all fossil fuels and switch to electric cars as quickly as possible to combat climate change. Others argue that you're trading one set of environmental issues for another due to the mining of minerals and metals used to make batteries. For those seeking to answer those questions, this book isn't for you.

Instead, the following chapters are based on the recognition that electric vehicles are here and will be coming in greater numbers. But more likely than not, battery electric vehicles will not be a one-size-fits answer to our diverse transportation needs.

While the issue of cars and the environment sorts itself out, the question remains whether an electric car is right for you. And it's a decision with many factors. How far do you commute? Does the place you park overnight or at work have access to electricity? Would an electric car be your primary or secondary vehicle? What about tax credits and incentives? Can you really afford one? What will it be like to own one? This book is designed for you to make an informed decision that's right for you and your wallet.

1. SHOULD I BUY AN ELECTRIC CAR?

There's no shortage of excitement around electric vehicles. You can't turn on your television without seeing a commercial for one. Manufacturers are gearing up to produce them in large numbers, with lots of media attention to this coming wave of electric cars and trucks. Governments are subsidizing their purchase and use, while other authorities are proposing bans on the sale of gasoline-powered vehicles. It seems the world is going mad for EVs.

Many of the vehicles getting all the attention, such as Tesla, the Ford Mustang Mach-E and F-150 Lightning, GMC Hummer EV and newcomers like Rivian and Lucid, seem more like luxury goods than transportation for regular folks. With average sticker prices well north of $50,000, these vehicles cost far more than the average car or truck, which, according to Kelley Blue Book, is hovering somewhere around $45,000. Not exactly the kind of bargain any self-respecting Tightwad would seek out.

Still, there are some more affordable electrics flying under the radar. These vehicles, like the Nissan LEAF, Chevrolet Bolt, Hyundai Kona and Kia Niro, are priced between $30,000 and $40,000. While still higher than comparable economy cars, some of these electrics can bring a smile a Tightwad's face thanks to tax credits and local incentives.

Before plunging headlong into the search for an affordable electric car, a shopper needs to understand how an EV works, the advantages and disadvantages of ownership and how to navigate this expanding automotive landscape to make the right choice at the right price.

The first step is determining if an electric vehicle fits your needs. Looking at rising gas prices, the attractive incentives and increasing availability of lower priced models, the simple answer might be a resounding yes. But range limitations, lack of access to an outlet for recharging, or the climate in which you live might argue against going electric. These are the factors to consider before taking the plunge.

An electric vehicle may be right for you if:

Your daily commute is less than 100 miles roundtrip.

An electric car would be a second vehicle in the household.

Where you live has a dedicated parking space with access to a 240-volt outlet or circuit.

Most of your driving will be in an urban or suburban environment with access to public charging.

You already have home solar panels.

The climate in which you live is temperate.

Owning an electric car is challenging if:

You rely on it as a primary vehicle and have no backup.

Your daily roundtrip commute is greater than 100 miles.

You live in an apartment or condo with no assigned parking or access to electricity.

You live in a rural area with little or no public charging infrastructure.

Spur of the moment road trips are part of your lifestyle.

The climate you live in has periods of extreme heat or cold.

For those looking to go electric on a budget, the least expensive models on the market today have somewhat limited range, usually between 100 and 260 miles. Don't pay attention to all the hype surrounding EVs that can go 300, 400 or even 500 miles on a charge. All of them cost upwards of $50,000.

How far you drive every day and whether you have access to a second vehicle that has some form of internal combustion are also factors. Road trips are possible with electric vehicles, but they take careful planning. Having a gas-powered alternative to use instead will take the range anxiety out of that spur of the moment weekend getaway.

Home charging is also a big issue. If you live in an apartment or condo, you might not have easy access to an outlet or won't be able to install a faster Level 2 charger. You can recharge an electric vehicle on an ordinary 120-volt outlet, but it could take more than a day to recharge an electric vehicle with the smallest of batteries. Having access to Level 2 charging at 240 volts will cut that time by as much as two-thirds.

Another consideration is whether you'll be operating your EV in an area where there's a good number of public charging stations.

These are quite handy if you have access to them while you're working or shopping. But be aware that public charging might cost more than double your home electric rates.

If you live in a rural area with long distances between population centers with public charging facilities, an EV is probably not the right choice for you.

Electric car ownership is attractive to those who already have solar power at home. That investment enables you to take advantage of lower cost electricity when charging at home. However, buying an affordable electric car doesn't justify the cost of adding an expensive solar system just to recharge it.

As part of the research for this book, I bought a 2022 Nissan LEAF S. In true Tightwad fashion, it is the least expensive electric car on the market today with a base price of about $28,000. It also has a somewhat limited range of nearly 150 miles. In the ensuing chapters, we'll walk through the shopping experience, ins and outs of rebates and tax credits, what it's like to drive and own an electric car, and what it costs to recharge, insure, and maintain it. Sharing these experiences will hopefully provide a roadmap for you to find, buy and enjoy an affordable electric vehicle.

Electric Cars are Nothing New

The first electric horseless carriage is believed to have been built by Scotsman Robert Anderson in the 1830s. However, the early batteries employed an irreversible chemical reaction that made them useless after they were fully discharged. It wasn't until the invention of the rechargeable battery by French physicist Gaston Planté in 1859 set the stage for electric cars of today.

Des Moines chemist William Morrison is credited with building the first American electric, a six-passenger wagon, in 1891. It was capable of a top speed of 14 miles an hour and helped create widespread interest in electric-powered transportation. By the turn of the century, New York boasted a fleet of 60 electric taxis.

The crude nature of the first gas and steam powered cars contributed to the early success of electrics. In addition to their lack of refinement, combustion engines were dirty and required considerable effort to crank start by hand. Steam engines could take as long as 45 minutes to build up enough pressure to get going. Electrics started instantly, ran nearly silently and, given that most were used in urban areas with paved streets, their limited range was not a big deal.

A 1901 Columbia is displayed at Pebble Beach with its charger.

In the early 1900s, electrics were popular, accounting for more than a third of all vehicles on the road, according to the U.S.

Department of Energy. They cost as little as $1,000, though many of these electric "town cars" cost upwards of $2,000. Underscoring this tech rivalry between electric and gasoline power was the fact that some manufacturers, like Studebaker, Pope and Oldsmobile, built both.

It's worth noting that in 1908, the year Henry Ford introduced the Model T, he bought a Detroit Electric for his wife Clara. At the time, electrics were viewed as "ladies' cars" because they were much easier to start and drive than their hand-crank internal combustion counterparts.

Electric vehicle production hit its high point in 1912, which was when, not coincidentally, gas-powered cars started using the electric starter invented by Charles Kettering. It was also the year that the $1,750 price of a typical electric roadster was undercut by Ford's mass-produced Model T which cost just $650.

Electric cars still retained a modicum of popularity, but their days were numbered. It was more than just the lower price of internal combustion cars that contributed to the demise of early EVs. Gasoline was plentiful and cheap, refueling took far less time than recharging, and a growing road network between cities and the demand for longer range made internal combustion vehicles better suited to an increasingly mobile American lifestyle.

Nonetheless, electric technology continued to evolve. A 1922 Millburn Light Electric 27L came with a battery pack mounted on rollers that allowed it to be quickly swapped out for a freshly charged array. It had a controller that allowed for four forward and two reverse speeds. When equipped with an 84-volt battery pack, the Millburn could travel about 100 miles between battery changes. That's impressive when you consider that 90 years later, the electric cars leading the EV comeback like the Nissan LEAF, Fiat 500e, Ford Focus EV and Volkswagen eGolf could muster only 80 to 90 miles on a charge.

This 1922 Millburn Light Electric was used by the White House.

But the factors that led to electric vehicles falling out of favor with consumers still exist today. Electric cars continue to cost more than their gasoline powered counterparts. Recharging them takes longer than a gas tank fill-up. Range is limited in comparison. The lack of rural charging infrastructure (beyond Tesla's Supercharging network) makes cross-country travel difficult.

Rebirth of the Electric Car

The renaissance for electric power in cars started in 1990. That's when California, which was allowed to set its own emission standards to combat smog, adopted a regulation requiring that by 1998, two percent of a major manufacturer's new cars sales in the state had to be zero emission vehicles (ZEV). That percentage would rachet up to 10 percent by 2003. The mandate gave rise to the first production EVs seen on American roads since 1935.

GM introduced the EV1 in 1996 in California and Arizona only.

The poster child for this first go around was GM's EV1, a two-seater that debuted with lead-acid batteries and a range of about 70 miles. A second generation with more advanced nickel-hydride batteries boosted that distance to 140 miles. GM, which provided the EV1 as a lease-only proposition, created a furor when it took the vehicles back and crushed them after California weakened its ZEV rules to allow non-electrics like natural gas and flex-fuel ethanol cars to count towards the mandate. Manufacturers also paid fines or generated ZEV credits through other means. Chrysler, for instance, offered the GEM neighborhood electric vehicle, which was a glorified golf cart, to avoid the penalties.

It wasn't until 2012 when the state passed a new round of initiatives reviving the ZEV mandate that the manufacturers began to produce the next generation of electric vehicles in earnest. The state also won a waiver from the federal government that allowed other states to follow its more stringent guidelines. A $7,500 federal tax credit was also enacted to spur EV sales. In anticipation of the new rules, Nissan was the first major manufacturer to offer a mass market electric vehicle in the LEAF when it went on sale in December of 2010.

The Nissan LEAF is the first modern mass market EV from a major make.

One year earlier, a small Silicon Valley startup, Tesla Motors, introduced an electric two-seat roadster. The company, founded in 2003 by Martin Eberhard and Mark Tarpenning, included backing by Elon Musk, who became chairman in 2008. Tesla introduced its full-size Model S sedan in 2012 followed by the Model X SUV in 2015. The Model 3 midsize sedan debuted in 2017 and Model Y midsize crossover came in 2020.

In addition to selling EVs, Tesla also benefited by selling the electric vehicle clean air credits other manufacturers needed to offset their sales of gasoline and diesel cars and trucks in the state. But beyond using those credits, the industry began a concerted effort to build electric vehicles to generate their own credits and to address the now growing demand for electric cars.

Slowly, the barriers to EV ownership are coming down with larger scale production leading to a reduction in vehicle costs. Advancements in battery technology enable faster charging capability and longer range. Improvements to the charging infrastructure, especially Tesla's national rollout of its Supercharging network, makes longer road trips feasible.

Cost Conundrum

While progress continues, in many cases electric car ownership is antithetical to Tightwad sensibility. That's because the day has yet to arrive where an electric vehicle without subsidies is cost competitive with its gasoline-powered counterpart. While simpler, (auto parts giant NAPA estimates that an electric vehicle has about 12 major moving parts versus 200 for a gas car), electrics still cost more to produce.

The big-ticket item here is the lithium-ion battery, which employs cathodes and anodes using precious metals to make its chemistry work. Strides are being made to lower costs through the use of less expensive metals and minerals. Further cost reductions will also result from large scale production. But the price differential between electric and gasoline power persists.

According to a white paper on electric vehicle production costs by the International Council on Clean Transportation, the average EV in 2025 will have a battery pack cost of about $8,000, with the electric drive module adding another $1,080. The report states that a conventional gasoline-powered drivetrain, which includes the transmission and exhaust system (with its expensive catalytic converter that also uses precious metals), costs $6,800.

The current disparity in pricing between gas and electric is evident in the 2023 Hyundai Kona, which comes in a choice of either traditional internal combustion or electric power. The gas-powered Kona in SE trim starts at $21,990, while a similar SE electric is more than $10,000 higher at $33,550. While the Tightwad in me flinches at that higher number, thanks to local tax breaks and incentives, that gap can be narrowed considerably. So, just because an electric vehicle carries a premium price tag, there are ways to bring that cost more in line with conventional cars.

But the subsidies may not last forever. In the case of the Kona, recent legislation makes it ineligible for the federal $7,500 tax credit because it's not built in North America. That same legislation reinstated the credit for General Motors and Tesla by removing a previous provision that limited the break to the first 200,000 units sold. But other language in the bill may take that credit away if companies don't source at least 40 percent of battery's material or manufacture at least half it here. So, for Tightwads looking to use tax credits to get into an EV have some major homework ahead of them.

Of course, the wild card is not just the purchase price of electric vs. gas but also operating costs. As recently as April 2020, the national average for a gallon of regular, according to the U.S. Energy Information Agency, was $1.87. Two years later that average had jumped to $3.67 per gallon. In California, motorists regularly see gas above $6 per gallon.

Using California as an extreme example, electricity rates for the

Golden State run about 23 cents per kWh. So recharging that 40 kWh LEAF should cost less than $10. Rounding up the range to an even 150 miles, the running cost of the electric is 6.6 cents per mile. Assuming you have a conventional gas car that matches the federal average fuel economy of 25 mpg and at $5 per gallon, covering that same distance runs 20 cents per mile. Note that both the gas prices and electricity rates are among the highest in the nation, so elsewhere the per mile running costs will be less. Still, even at $2.50 a gallon or 10 cents per mile, the math still favors an electric vehicle.

Even though carmakers are looking to continually lower their manufacturing costs, it doesn't mean vehicles, both gas and electric, are going to get less expensive as time goes on. New electronic driver aids and safety assists are expected by consumers, so the average price of vehicles continues to climb.

How to Beat the Market

Despite all this upward pressure in prices, you can still find a bargain in a market plagued by vehicle shortages, a lack of manufacturer incentives and an environment where buyers, on average, are paying hundreds, if not thousands, over sticker prices. The strategy for Tightwads is simple: You shop for vehicles that are not in high demand or, conversely, shop in segments where there is a glut of supply.

Affordable electric vehicles hit the sweet spot on both counts. While demand for new vehicles currently outstrips the supply and inventories are at record lows, both which fuel the upward price spiral, not all new car segments face the same challenge. The hot tickets right now are crossover SUVs, pickups and even minivans.

Not only are the number of vehicles available limited, but also there are fewer incentives. Buyers are either paying more than list price or not finding exactly the model that meets their needs. On the flip side, sales of traditional sedans and compact cars continue to fall in the face of the popularity of trucks and SUVs. The few incentives available are on these models. However, the choices here are becoming more limited as manufacturers exit midsize and compact segments in favor of small and midsize crossovers.

Here's where EVs are becoming more attractive. While there is a lot of buzz surrounding electric cars, that attention far outstrips the actual sales numbers.

Until recently, electrics have accounted for as little as two percent of overall sales in a market that had topped the 17 million mark for three years running before the COVID pandemic hit. While overall sales are down because of production shutdowns and chip shortages, the EV share of sales, thanks to a slew of new models, has doubled to four percent.

While there is considerable interest in more expensive EVs with sold-out launches of higher-priced models like the Ford Mustang Mach-E, the needle isn't moving much on lower priced models like the Nissan LEAF and Hyundai Kona, at least until fuel prices began to spike.

 Still, manufacturers have huge incentives to build and sell these vehicles to generate the credits they need to sell their large and profitable gas-powered pickup trucks and full-size SUVs.

It's the incentives that you don't see that can make the difference in coming out ahead by buying an electric vehicle. The manufacturer might be willing to make less money or even take a bit of a loss moving a lower priced EV so that they can use the fuel economy credits generated by that sale to make more money on a less efficient but higher priced and higher demand SUV, pickup truck or luxury car.

Evidence that auto makers have built flexibility into their pricing structures around the federal tax credit can be seen in GM's retail price of the Chevrolet Bolt. When the Bolt was still eligible for federal tax credits, the least expensive model cost $37,495 including $995 delivery. If you qualified for the full $7,500 credit, the effective out-of-pocket expense dipped below $30,000. A 2022 Chevrolet Bolt, no longer eligible for the federal tax credit, carries a sticker price of $31,500.

So, what are buyers willing to pay for electric vehicles? A 2020 study conducted by engine oil marketer Castrol found that consumers are looking for a price point of about $36,000 for a new electric car.

The study says, "Price is the number one priority for consumers in the U.S., with 57 percent of those surveyed saying that EVs are currently beyond their budget." The study also found that misconceptions about maintenance costs could be stopping consumers making the switch: 65 percent of U.S. consumers say that these costs were preventing them from buying a fully electric car. This suggests that many are unaware that the average cost of ownership of an EV over its lifetime tends to be lower than an internal combustion engine vehicle.

The study's respondents expect a vehicle that takes only 30 minutes to recharge. That is a pretty tall order given the current state of technology that sees electrics with a modest range taking about seven hours on a home Level 2 240-volt setup. While 60 percent said they are taking a wait-and-see attitude towards electric vehicles, a substantial number said they would be interesting in buying one as early as 2025.

Even though the survey points to expectations at odds with the current reality of the electric car market, there are opportunities for the average buyer to find an affordable EV, thanks to the growing availability of lower priced options and existing incentive programs. But the question that looms large in any discussion about electric cars, is how far can it go?

Tightwad Pro Tip:

Given the disparity in gas prices across the U.S., buyers have a better chance of finding affordable deals in less densely populated areas with lower gas prices. In states like California, with prices spiking up over $6 per gallon, don't expect to pay sticker or below until there's a decline in the cost of gas.

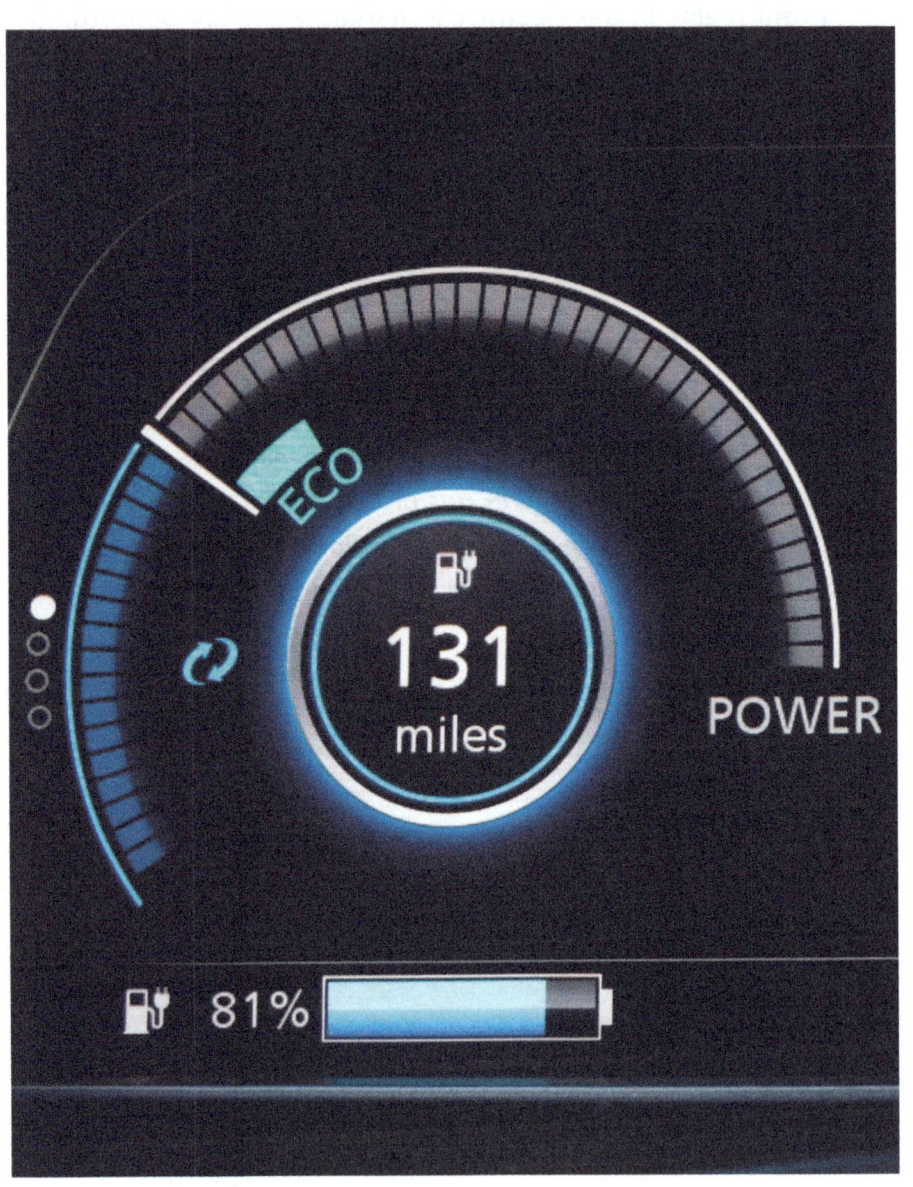

2. HOMING IN ON THE RANGE

Range is a huge factor when considering an electric car. The question of how far you want to go on a charge is a function of how much you want to spend. It's akin to basing the purchase of a gas car on the size of its fuel tank. Since the battery is the largest single cost component on an electric vehicle, you can pay a lot more for the same car right down to model, trim level and

2022 Nissan LEAF S

features mainly due to the size of the battery.

The 2022 Nissan LEAF S with the 40-kWh battery, which gives it an EPA range of 149 miles, starts at $27,400. A comparable LEAF S Plus with a 62-kWh battery and a 225-mile range, costs $32,400, a $5,000 difference.

From a Tightwad's perspective, the question is whether that extra 75 miles is worth the cost. Another factor is that a larger battery may take you farther between charges, but all things being equal, it could take a lot longer to recharge than a smaller battery pack with shorter range.

The Range You Need

When looking at electric car affordability, the key question isn't how far you *want* to go on a charge, but rather, how far do you *need* to go. Also framing that discussion is whether your electric vehicle is your primary daily driver or plays a secondary role in the family fleet. If it's the former, a longer range may be necessary. If it's the latter, you're likely not to need as much.

According to a recent Department of Census report, the average commute in 2019 took an average of 27.6 minutes each way. It covered about 30 miles round trip. In a 2017 National Household Travel Survey, the U.S. Bureau of Transportation Statistics found that Americans travel an average of 14,500 miles per year by car, which works out to 40 miles per day.

That 40 miles per day expressed as a weekly number means that if you want to charge only once a week, an electric vehicle should have a range of at least 280 miles. That number isn't far off consumer sentiment in that Castrol electric vehicle study mentioned earlier. Survey respondents said an average of 319 miles between charges was the tipping point to where they would consider buying an electric.

This 300-mile-plus range expectation is roughly what many consumers get from a tank of gas. And most drivers usually fill up once a week. The survey underscores the fact that people expect roughly the same level of range and performance from electric vehicles that would allow them to directly replace their current gas-powered vehicles.

However, if you recharge your electric car at home overnight two or three times a week rather than making the usual weekly trip to the gas station, you might find a shorter-range EV an acceptable trade-off versus a more expensive model with more range.

You also may want to consider the nature of your commute. Electric vehicles are more efficient in stop-and-go city driving than steady highway cruising. The lower speeds combined with the fact that an EV doesn't draw power from the battery when it's stopped is reflected in the difference between the EPA's higher city and lower highway mileage for electrics and most hybrids. The opposite is true for gas cars, which are more efficient at highway speeds than in stop-and-go traffic.

So, a 30-mile highway-speed commute at higher speeds in an electric vehicle may end up using more range faster than a longer trip in stop-and-go city traffic.

Just as early electric vehicles found their place primarily as town cars, most affordable electrics today are perfectly comfortable in that same role. An EV makes an ideal train station car for those who commute to city centers, or as an urban runabout. As a second vehicle in the household, they are also well-suited for suburban errand running, short-range commutes and day or overnight trips of 100 miles or less.

The operational distances are likely to improve over time as the recharging infrastructure, particularly kiosks with fast charging capability, expands. Cross-country trips in an EV that gets at best 200 miles on a charge will continue to be a challenge. When it comes to the convenience factor, the ability to quickly refuel remains the internal combustion vehicle's trump card. As Tightwads know, time is money.

Range Stealers

Once you've determined what kind of range you'd like from your electric vehicle, you also need to be aware of the wide variety of factors that can help and hurt your ability to cover that distance. We all know that a lead foot can take a bite out of your gas car's ability to deliver its EPA-certified mpg ratings. We've all heard the term "Your mileage may vary" and that saying applies in spades to electric cars.

Knowing the remaining range is critical information.

Electric cars have unique characteristics, like regenerative braking, that can improve your range. While it lacks the visceral excitement of street racing, hypermiling is a time-honored Tightwad tradition that boosts the mpg in gas and hybrid cars by avoiding aggressive acceleration and heavy, last-minute braking. These same light foot techniques that work so well in conventional cars also can add miles between charges on electrics. Gentle acceleration reduces the draw on the battery, while coasting whenever possible and using the motor's regenerative braking to slow the vehicle returns energy to the battery pack.

Conversely, the instant torque and neck-snapping acceleration of electric power is intoxicating. If you give in to this adrenalin rush with jack rabbit starts and foot-to-the-floor passing maneuvers, you'll quickly suck the life out of your battery pack. Tesla points with great pride to its 2.0 second 0-60 mph acceleration of its Model S Plaid. Heavy-footed driving costs range, perhaps as much as 20 percent.

Just toggling between modes in most electrics, many of which have eco, normal and performance settings, will give you an idea of how much it will cost you in the estimated range displayed on the dash. These driving modes determine how quickly the motor responds to the accelerator as well as managing things like the air conditioning and heating system. It also demonstrates how vulnerable range is to not only how fast your drive, but also the accessory load. It turns out that the latter is more critical in an electric than a gas car.

These ancillary systems, from lights and audio to heating and air conditioning, use energy. In a gas car, the engine has an easier time dealing with these so-called parasitic losses. While the engine's main function is to drive the car, its mechanical operation allows it to also drive things like the water pump used in the cooling system, the air conditioning compressor and the alternator. Most of these systems have little effect on overall fuel economy, except for air conditioning, which the EPA says has an impact ranging between 5 to 20 percent.

In an electric vehicle, the motor has only two functions: driving the car or putting charge back into the battery. All the other accessories, including heating and air conditioning, are powered by the battery pack and/or the auxiliary 12-volt battery. When it comes to heating the cabin, a gas car has a distinct advantage in being able to use the hot engine coolant as a heat source for the cabin. Not so on an electric vehicle. Remember that 100-mile range 1922 Millburn Light Electric? It carried an on-board coal-fired heater for passenger comfort.

According to the Society of Automotive Engineers, a battery electric vehicle can lose as much as half to 60 percent of range from cabin heating, ventilation and air conditioning (HVAC). And while many of the electric vehicles today tout systems that allow them to pre-heat and pre-cool the cabin before you begin your journey, keep in mind that that pre-conditioning an unplugged car will cost range.

If you live in a colder climate, you may want an EV that uses a heat pump instead of an electric resistance heater. Heat pumps are usually found on higher trim levels or more expensive EVs. Utilizing a coolant medium, these heat exchange units are used to provide both cool and warm air to the cabin. These systems are as much as three times more efficient as an electric heater.

The battery also reacts to extremes in weather, which also affects range. According to data compiled by Geotab, a fleet management specialist, electric vehicles on average see their rated range drop as much as 54 percent in 15 degrees Fahrenheit weather. At an optimal 70 degrees, range peaks at 115 percent of its rating, while in 110-degree heat it may lose around 35 percent of its range.

Charging Ahead

How often and how you recharge your vehicle can also influence range and battery life. The ideal state of charge for most batteries is between 20 and 80 percent. Topping off a charge to 100 percent or routinely running your battery as close to zero as possible can shorten its life and compromise its range. Battery life is also expressed in the number of charge and discharge cycles. Fast Direct Current charging rather than AC Level 2 charging may impact battery life and range, although recent studies have put the degradation down to around five percent.

Other factors that reduce range are aerodynamics and rolling resistance. While you can't do much to change the shape of your car to cheat the wind, you can avoid losing range by keeping extraneous bike and luggage racks, flags and spoilers (no matter how cool they look) off your car's body. Your tires can have an impact on how far your vehicle travels on a charge. Underinflated tires result in higher rolling resistance that must be overcome. Maintaining proper tire inflation is a key to maximizing range and performance.

How much cargo and the number of passengers on board also will have an impact on range. If you routinely carry a lot of luggage or three or four passengers, that extra weight will require more power and consequently, more frequent charging. Only take the things along on your trip you need and keep the golf clubs, bowling balls and anvils in the garage.

As for towing? Forget about it. Pulling any kind of trailer behind you will destroy whatever mileage gains you think you can get by hypermiling or running in sweltering heat without the air conditioning on. While the muscular torque of electric vehicles in general would make them tow friendly, the added weight and rolling resistance of even the lightest camper trailer costs as much as 25 to 30 percent of an electric car's range.

Keeping range in perspective is the difference between being anxious or confident in your electric car purchase. Understanding the advantages and disadvantages of EV technology will align your expectations with how these vehicles function in the real world. They can't do the same things as your old car, especially when it comes to the convenience of fast refueling and the ability to take spontaneous road trips. But there are other attributes in driving ease, the convenience of home charging and minimal maintenance that favor the electric car.

The range question as part of the affordability equation will continue to be an issue even though strides are being made in lowering the kWh costs of battery packs. The first generation of affordable electric cars at best boast a range between 90 and 100 miles. Improvements to the battery packs of some of these vehicles extended the range out to about 150 miles. Besides the LEAF, which offers both 149 and 225-mile versions, other low-cost electrics like the Chevrolet Bolt, Hyundai Kona and Kia Niro can go about 250 miles between charges. If you're looking for 300 miles plus (approximately the distance you'll go on an average tank of gas), you'll find prices above the $40,000 threshold. We will look at all these options in Chapter 6.

Range isn't the only major difference between a traditional gas power and electric cars. Knowing how an EV works is the first step to understanding what to expect from driving and owning one.

Tightwad Pro Tip

If your EV is equipped with heated/ventilated seats, use those first before turning on the main HVAC system in hot or cold weather. You'll find these accessories will have little or no effect on overall range.

3. HOW AN ELECTRIC CAR WORKS

It's magic. At least it seems that way, especially to car nuts with rudimentary knowledge of how an internal combustion works. A gasoline engine ignites a highly volatile fuel mixture in a cylinder to push down on a piston. That mechanical force is harnessed by a transmission that sends it to the wheels. Except for the electrical spark, it's purely a mechanical operation.

Electric cars are different. At its core is a large reservoir of electricity in the battery pack that spins electric motors to provide propulsion. The heart of a conventional car is its engine; in an electric, it's the battery.

Batteries trace their origins back to Alessandro Volta, an Italian scientist, who created the voltaic pile, a rather inelegant description of a collection of zinc and copper separated by a medium soaked in acid.

This amalgamation creates a sustainable electric charge. He is also responsible for Volta's pistol, an invention that tests the properties of methane gas. It allowed for the ignition of a flammable gas in a closed environment, which some believe set the stage for the internal combustion engine.

While Volta's experiments resulted in the first batteries, his process found limited application until French physicist Gaston Planté figured out a way to create a rechargeable battery, and thus the modern-day concept of range anxiety was born.

At the time Planté began his work, the best available battery based on the voltaic pile developed by Volta was called the Daniell Cell. However, its irreversible chemistry and low output made it impractical for widespread application. In 1859, Planté created a battery that employed positive cathode and negative anode electrodes. The cathode was made of lead, while the anode lead oxide. When immersed in sulfuric acid, the electrons stripped from the anode would create an electric charge as it traveled to the positive cathode. This chemical reaction wasn't just a one-way transfer. If an electrical charge was attached to the cathode, the process would be reversed, recharging the battery.

Each cell would produce about two volts of power and when six were combined created a 12-volt lead acid battery. While the technology has been refined to increase battery life by advancements in the electrode design, the same technology developed by Planté in the middle of the 19th Century still finds widespread application in today's automobiles.

However, lead-acid battery technology can only be scaled up so far. To power an electric vehicle, lead acid proved to be heavy and not particularly energy dense. The original electric vehicles of the early 20th Century using lead-acid batteries at best had a range of 100 miles. As you'll recall, GM's EV1, the two-seat commuter car introduced in 1996, could go only 70 miles between charges using lead-acid technology. Battery chemistry needed change. At first, it was a nickel metal hydride electrolyte. Used in the EV1, range doubled to 140 miles. Now, lighter weight and more energy dense lithium-ion batteries are current state of the art technology.

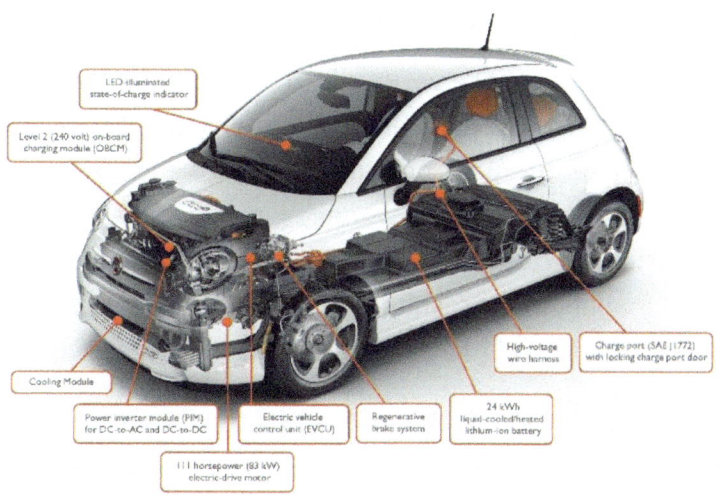

A cutaway of a Fiat 500e shows the major components of an electric car.

According to the International Research Journal of Engineering and Technology (IRJET), "Lithium-ion technology today represents the best compromise between capacity, volume and mass in the electric mobility sector." Lithium-ion batteries provide five times the energy density of nickel-metal hydride batteries in a lighter package. They also can take more recharging cycles before battery life is affected. Manufacturers routinely have 8-year, 100,000-mile warranties on their battery packs that include replacement if they fall below 70 to 80 percent peak efficiency during that period.

Lithium-ion batteries are popular because the charge doesn't dissipate over time, they are quicker to recharge than older battery chemistries and they're virtually maintenance free.

The battery pack needs a thermal management system to keep it from overheating. In less expensive cars, the batteries are air-cooled; in more expensive models, there's liquid cooling. The latter is a more active approach where a coolant is circulated in and around the battery pack to prevent the electrolyte solution from overheating when the battery is charging or discharging.

Cell configuration can take various forms. In a Tesla, there are 7,104 of them. Called 18650 cells, each one is about the size of a AA battery. The entire pack provide 85 kWh of energy. Using so many cells is costly, but the advantage lies in being able to route cooling channels to individual cells. This more advanced approach to thermal management results in a longer range, better fast-charging capability and high-power output for quick acceleration.

The Nissan LEAF uses cells that are laminated layers with no active liquid cooling. Manufacturing is simpler and results in a compact size and lower cost. The drawback is a lower energy density and limited range.

General Motors' next generation of electric vehicles will use its new Ultium battery technology which features 24-cell modules

GM's Ultium battery pack

that are about two feet long and four inches thick. Each three-pound cell is enclosed in a nearly half-inch-thick envelope or pouch. GM plans to package these cells in units ranging from a 50 kWh with 144 cells up to 200 kWh with 576 cells. The latter is good for a range of 450 miles.

Putting Battery Current to Work

Now, how do you get that energy from the battery to the wheels? An electric car uses a device called an inverter to change the electricity, which comes out of the battery as direct current, into alternating current used to drive the electric motor.

Also incorporated in this circuitry is a converter which either steps up or steps down the voltage from the battery. Stepping up the voltage is required by the traction motor, and stepping it down allows it to be used for other systems like heating and air conditioning.

Think of the inverter as the controller for the accelerator. It modulates the power coming from the battery to vary the speed of the electric motor and, in turn, the vehicle's velocity.

A basic electric vehicle will have one motor driving either the front or rear axle that relies on charged magnets to create rotational force.

The stationary housing of the electric motor, called the stator, uses copper windings and magnets to create an electromagnetic field from the battery's converted AC current. That force field causes the rotor inside housing to spin, providing the rotational mechanical power to turn the wheels.

The copper wire and electromagnets are the most expensive parts of an electric motor. According to the International Energy Agency, over a mile of copper wire can go into the windings of a typical motor. The average electric vehicle uses 183 pounds of the metal compared to 18 to 49 pounds employed a conventional car.

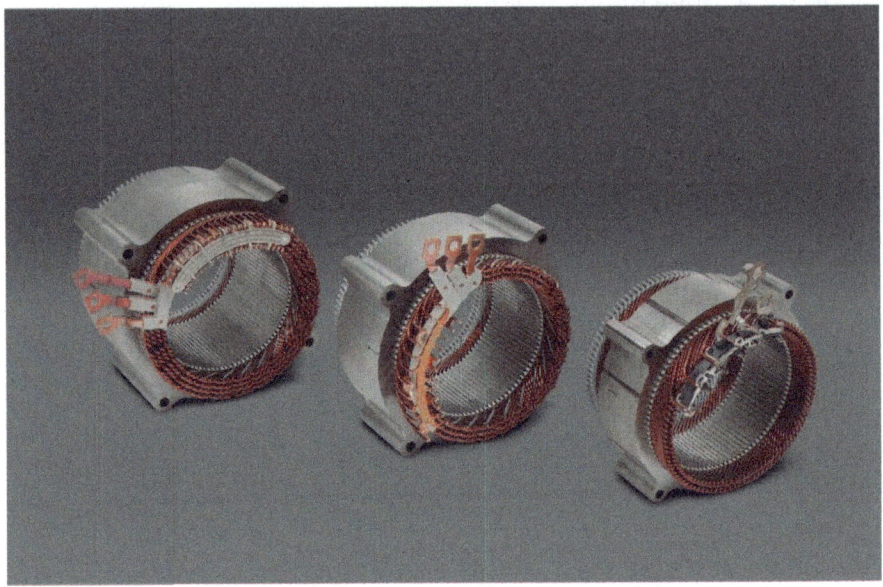

GM uses three different stator sizes for its Ultium electric motors.

The permanent magnets incorporate two rare earth minerals. One is neodymium, which adds strength, the other, dysprosium, increases heat resistance. Both are needed to protect the magnets from the high temperatures and forces generated by a rotor that can spin up to 20,000 rpm.

While an electric car has fewer components, the cost, as we'll see with the batteries, can be high from the use of these rare earth minerals and metals. That's not to say gasoline-powered vehicles are immune to similar expense. Even as car companies have lowered manufacturing costs, the catalytic converters used to clean up their emissions also rely on expensive metals like platinum and palladium, which makes them attractive to thieves.

As the source of the vehicle's motivation, the electric motor is analogous to the engine in an internal combustion engine. Its output, however, is measured in kilowatts. The LEAF S, for example, employs a single electric motor driving the front axle and is rated at 110 kilowatts. That output is the equivalent of 134 horsepower, and its torque rating is 236 pound-feet (lb-ft).

In simplified terms, horsepower is a measure of how much work can be done over time, while torque is a measure of power delivered at a given moment. Horsepower determines how fast a vehicle can go, while torque is a measure of how quickly it can reach that speed. An electric motor produces peak torque instantly, while an internal combustion engine takes more time to spin up to its maximum output. As a rule, gas-powered cars are capable of higher top speeds, but electric vehicles, with the instant delivery of torque, feel quicker off the line.

The Regenerative Braking Advantage

The interesting characteristic of an electric motor is its ability to generate current that can be sent back to the battery for storage. This ability to reverse the flow of electrons, which is called regenerative braking, converts a vehicle's kinetic energy to reusable power while helping to slow the vehicle down. Internal combustion engines also can use the engine for braking, essentially by using a lower gear and the forces of the engine itself to slow the vehicle. The difference here is that engine braking doesn't put at any gasoline in the tank, while regen braking helps recharge the battery.

Unlike a conventional car which connects the engine to a transmission with multiple gears, electric motors employ a single-step gear reduction to better match its high-speed spinning to the lower rotational forces needed to drive the car. There's also a differential gear split that sends power to the wheels at each end of the axle. Because an electric motor operates over such a wide power band, there's no need for a traditional stepped-gear transmission used in conventional cars. Also, there's no need for a reverse gear – backing up is just a matter of making the motor spin in the opposite direction.

Honda's Clarity EV uses a single motor on the front axle. The inverter and electronic controls are stacked on top.

More affordable electric cars have just one electric motor usually positioned on the front axle for packaging purposes. Since most new vehicle assembly is geared towards front-wheel drive, virtually all low-cost EVs on the market are configured that way.

Some higher performance EVs offer rear motor placement. Rear-drive cars provide more spirited handling by splitting up the steering and driving duties between the respective front and rear axles. All-wheel-drive electric cars are equipped with front and rear electric motors. As you'd expect, doubling the motors comes at a much higher price.

With just a battery pack, inverter/converter and motor, an electric car in its basic form is much simpler than a gasoline-powered one. The main difference between electric and gas vehicles is what's making the wheels turn. An electric car is like a personal computer while a conventional car is more of a manual typewriter. Both can produce printed documents, but one uses electrons while the other requires mechanical action to put the letters on a page.

Beyond its different means of propulsion, an EV still needs to do all the things that a conventional car does, like steering, stopping, and keeping its occupants comfortable, connected and safe.

Most of the systems on a conventional car, like the heating and air conditioning, power steering, brakes and electronics, are run off the engine. The air conditioning compressor, alternator, power steering and water pumps are belt driven. Also, the heating system uses waste heat from engine cooling to warm the cabin.

Electric vehicles also rely on hydraulic brakes and steering (with electric assists), on-board electronics and climate controls. But rather than being driven by the electric motor, most of these functions rely on the battery pack for their operation.

As mentioned earlier, a little-known aspect of electric cars is that they also depend on a separate 12-volt system powered by a conventional lead-acid battery to run things like the exterior and interior lights, infotainment, power seat adjustments, heated steering wheels and seat heating/ventilation. This separate battery is kept in a state of charge by a combination of current from the electric motor's regenerative braking and the battery.

Larger systems that require more energy to run, like cabin heating and air conditioning, draw their power directly from the battery pack. You can see this impact on most electric cars by simply turning on the air conditioning and watching the instrument cluster's estimated range reading drop.

Less expensive electric cars will use air conditioning compressors and resistance heaters for climate control that can put a big load on the battery when run at high levels. Higher priced models will use heat pump technology that has less of an impact on range. It's a good question to ask if the EV you're interested in has one or if it's an available option.

Now that we've covered the basics of how an electric car works, let's get behind the wheel.

Tightwad Pro Tip:

When considering an affordable electric vehicle with a single motor, look for a front-drive model over rear-drive for its better traction in bad weather.

4. WHAT IT'S LIKE TO DRIVE AN ELECTRIC CAR

Say you've never driven an electric car and if by magic you were plopped down behind the wheel while it's cruising down the freeway. You'll find there's would be little to distinguish the experience from that of a gas car.

At those speeds, most conventional drivetrain noises are muted or drowned out by tire or wind noise. The steering will feel the same, you might be listening to the radio, or have air-conditioning wafting from the vents. There is little, other than the instruments displaying range or state of battery charge, that would appear to be out of the ordinary.

Still, there are stark differences in the way that electric and traditional cars operate. In some cases, these characteristics are easy to spot, while others are more nuanced. Understanding how an electric car responds is critical in setting your expectations of the ownership experience.

Let's start with the obvious. Power. Mash your right foot down and it doesn't matter if you're in this Tightwad's choice, a Nissan LEAF S, or a Tesla, there is an immediate response. Depending on the power of the electric motor, especially in the higher dollar dual-motor hot rods, the term neck-snapping acceleration becomes all too real. In fact, one of the fastest cars on the planet is the Plaid performance version of the Tesla Model S, said to be able to hit that 60-mph benchmark in just under two seconds.

While it may not be in the Tightwad's wheelhouse to spring for a Tesla Model S, the more affordable Nissan LEAF can accelerate to 30 mph is about three seconds. In fact, many electric vehicles are capable of that performance because of the instant response of the electric motor at full torque when you push down on the accelerator.

Most gas-powered vehicles may take as much as a half-second longer to get to 30 mph than their electric counterparts because of the time it takes for an engine to spool up to deliver maximum torque and horsepower. A half-second may not seem significant, but that immediate response reinforces the feel that the car is ready to go when you are.

What an electric car delivers at immediate response, its acceleration curve quickly hits its plateau and continues to gain momentum more slowly from there. While gas cars, especially high-performance models, can accelerate almost as quickly, they have more pulling power as the engine's revs (revolutions per minute) climb.

Why is this important? For many that 0-60 mph equates to confidence in accelerating up to freeway speeds to be able to smoothly blend in with traffic. However, in city driving, a case can be made that an EV inspires confidence with quicker low-speed acceleration when darting in and out of traffic. Just as an electric car delivers its best mileage in a stop-and-go traffic, it's also satisfying to drive for its quiet, nimble and quick nature in the hurly-burly of an urban setting.

The quick response of an electric motor also offers benefits in mid-range passing maneuvers where you need an instant burst of power. A traditional car may pause a beat as it either spins faster or kicks down into a lower gear to give it that extra push. A January 2020 *Motor Trend* comparison of a Kona EV vs. Kona with the more powerful 1.6-liter turbocharged engine saw the EV take 3.0 seconds to go from 45 to 60 mph versus 3.3 seconds for the gas model. Again, that difference may seem negligible at face value, but the instantaneous response to your right foot is reassuring when making a move.

Give Me a Brake

As for braking, the difference between gas and electric vehicles is easily discerned. A gas-powered car will have some engine braking when you lift off the gas, but most of the stopping power comes from how hard and quickly you push on the brake pedal. Additional braking come from the downshifting of the transmission into lower gear ratios, although this mechanically induced slowing is often subtle.

Braking in an electric car is much more immediate. Thanks to regenerative braking, the motor itself provides the first bite. How aggressive the motor is in slowing the car is adjustable, either through selecting a drive mode or by using paddles on the steering wheel that allow you to dial in more resistance from regen braking.

The Nissan LEAF has several ways to boost the effects of regen braking. In addition to Normal and Eco modes, the last of which produces more of this effect, you can also select a B or Brake mode in the gearshift that also increases regen braking.

The ePedal switch enables one-pedal driving.

The highest level, however, comes from a switch marked ePedal that allows "one-pedal" driving. This mode combines regenerative forces and the standard brake system to help bring the car to a complete stop when you take your foot off the gas. It doesn't require application of the brake pedal. You modulate the braking distance by how quickly or slowly you ease off the accelerator.

There is a bit of a learning curve involved in one-pedal driving, but it's a skill that easily mastered, especially if you're accustomed to driving a golf cart. You also must be prepared to use the traditional brake pedal if someone cuts in front of you or a quicker stop is required.

So, how do you let people know you're braking with one-pedal driving since the brake lights are typically activated by the brake pedal? When the ePedal mode is engaged, the brake lights illuminate as soon as you back off the accelerator. And they stay on as long as you're not pushing down on the pedal and will remain on while you're stopped. They go off once you press down on the go pedal.

Other electric vehicles use steering wheel paddles to modulate regenerative braking. You can pull on a paddle and use it like a hand brake to bring a vehicle to a stop, or in other scenarios, a two-paddle setup is used, one side increases regenerative braking and the other reduces it. In cars with adjustable regenerative braking, the highest level usually incorporates some form of one-pedal operation.

This heavy reliance on the electric motor to slow the car gives EVs a distinctly different feel when braking. A gasoline car relies primarily on its hydraulic brakes, with engine braking playing a minor role. As a result, the brake feel (the amount of effort needed on the pedal to bring the car to a halt) is often consistent, linear and predictable.

Brake feel on an electric can be more like a relationship on Facebook tagged "It's complicated." While it is possible to dial in the level of regenerative braking, the battery's state of charge also plays a role in how aggressive those forces are, especially when the vehicle is left in a normal mode.

Being able to balance regenerative and traditional hydraulic braking action in an electric vehicle (and in most hybrids as well) is at best a black art. The algorithms controlling these curves are getting better, but they are not infallible. A battery in a low state of charge is receptive to taking as much regen braking current that the motor can generate. A battery closer to full charge, not so much. Figuring out how much regen and hydraulic braking is required is more consistent when the one-pedal mode is engaged because the car is, in essence, talking to itself.

But in normal driving when you need to apply the brake to bring the car to a halt, the amount of pedal pressure from the driver adds an unknown variable to the equation. You may find yourself applying too much brake and stopping short if the system is demanding a lot of regen braking due to a lower state of charge. Or you may be too light on the initial application of the brake and end up jamming on them because of a lighter-than-expected regen braking that results from a battery closer to full charge.

Braking feel is just one calibration that's specific to each model. Steering precision and weight as well as the responsiveness of the accelerator can vary tremendously. Consequently, when shopping for an electric car it pays to test drive a wide range of vehicles from different manufacturers in your price range to find out the best one that suits your driving style.

Take the Wheel, Please

Virtually all the controls are similar in electric and gas cars. You have a steering wheel, accelerator and brake pedals, and an electric parking brake. There are either capacitive touch controls or traditional analog knobs for things like the audio system and heating/air conditioning. You can get the same features such as heated and cooled seats, cruise control, and powered seats, mirrors and windows. All are very familiar and easy to use.

Since electric cars don't have a traditional geared transmission, instead of a shift lever, there's a variety of approaches-in selecting drive, reverse and park. Again, this is not dissimilar to traditional gas cars that also have push buttons, rotary knobs or steering column stalks to select the gears.

There are also buttons for driving mode, like ECO, which remaps the accelerator to provide more leisurely (and energy saving) acceleration. The Normal modes provides quick off-the-line performance but that comes at the expense of range. There's also a One-Pedal driving mode button that engages that feature.

Because of the relatively silent nature of electrics, the National Highway Traffic Safety Administration requires them to make some sort of noise at low speed. Going forward, most of them make a low pitched echoey sound that mimics what you would hear coming from flying saucers in cheesy 1950s sci-fi movies.

When backing up, you'll get a louder chime or beeping tone. The LEAF makes a noise not unlike a naval destroyer using sonar to ping for enemy subs that might be lurking beneath the surface of the driveway. Others may have more pleasant chime tones in reverse. When test driving an electric car, stand outside and have the salesperson back it up for you to see how innocuous or annoying that sound will be. Your neighbors will thank you, especially if you leave for work early in the morning.

As mentioned in our earlier discussion about range in Chapter 2, accessories and climate controls play a big factor in how far your EV will go on a single charge. When taking the test drive, be sure to check the "miles remaining" readout on the instrument panel. Turn on and off both the air conditioning and heating to levels you'd normally use to see the effect on remaining range. Things like the audio system and lights won't affect range since there's that separate 12-volt system with its own lead-acid battery to power these items.

Then there is weight. If you're attuned to vehicle dynamics, you may notice that your electric may feel heavy. It is, thanks to the battery pack. A great head-to-head comparison is the Hyundai Kona SEL. The front-drive gas version tips the scales just under 2,900 pounds, while the Kona EV comes in at 3,715 pounds. Most battery packs add 800 to 1,200 pounds to the average EV over a comparable gas-powered car. While an electric car may have a heavier feel, that weight is located beneath the cabin, which results in a lower center of gravity and a less tippy feel.

Smooth Operator

Thanks to power assists on the steering, a vehicle like the Nissan LEAF feels nimble. The action is light, yet precise. While there is some tire and wind noise at speed, there's no discernible noise or vibration from the powertrain itself.

It's this smooth, almost effortless driving that sets the EV experience apart. Those who feel driving is a chore may find driving an electric vehicle easier and less stressful. On the other hand, those passionate about internal combustion driving-may feel driving an electric is a bit sterile. Both points of view have merit – it boils down to what you expect from your car.

Ownership goes far beyond the rational need of getting from Point A to Point B. There are plenty of emotional reasons why you want a particular car that often trump the rational. When evaluating whether an electric is right for you, be clear on the reasons why you want one.

On the rational side of the ledger there's an element of practicality to an electric vehicle. It can be used to run errands, drive to work and meet basic transportation needs without having to visit a gas station. There are the environmental reasons for using electricity instead of fossil fuels (although where your electricity comes from is also part of the equation of the greenness of your purchase). Rational reasons aside, there's also the cool factor of being an early adopter of a leading-edge technology by going electric.

If affordability is key to your decision, you must keep in mind the limitations. You won't be buying an electric with the longest range, nor will it be the fastest one on the market. The ability of an affordable EV to take long trips, haul stuff or tow toys for use on extended recreational outings will be limited. While there are long-range electric pickups, off-road vehicles and even RVs coming, they are-more expensive luxury items than practical alternatives to less expensive gas pickups and SUVs. For the Tightwad, an EV purchase at this stage of the game is more about low-cost daily transportation.

For those naysayers who believe that electric cars are too

expensive, aren't practical and won't be sold in any significant numbers, they haven't really looked at what's on the market today. Nor are they aware of the positives that going electric have for the everyday car buyer.

The question is finding the right answer that addresses not only your practical pocketbook concerns, but also those emotional factors that make electric car ownership special.

Tightwad Pro Tip:

A test drive is an absolute must when shopping for an electric car. In addition to getting some real-world seat time, it's an opportunity to learn about the features, accessories and systems on the vehicle that affect range and the driving experience.

5. HOW TO BUY AN AFFORDABLE EV

Here's where the rubber meets the road. Just as EVs are changing how we think about cars, the buying process itself is also evolving. While nothing can replace seeing a car in person and taking it for a test drive (both of which you should do before buying any kind of car), the internet empowers you to do most of the legwork from home.

Doing the necessary research online is critical to getting the best deal. New digital tools now available enable to you shop for the right vehicle, determine a proper strategy for dealing with your trade-in if you have one, allow you to qualify for financing, and arrange for insurance well before you step foot on a dealer's lot.

You also can find and negotiate prices from different dealers, discover what type of manufacturer incentives are available and determine the eligibility of your purchase for federal tax credits, state and local rebates, utility rate breaks and car-pool lane access, if applicable in your area.

Keep in mind that due to the patchwork of regulations with some states following California's stricter emissions regulations, some electric vehicles may not be offered in a state that adheres only to federal standards.

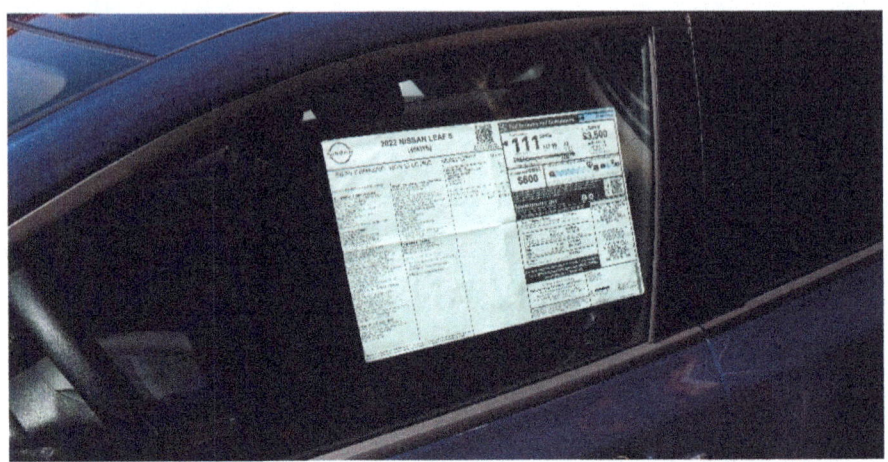

The window sticker offers information on range and operating costs.

Let's start with the lowest priced electric car on the market, the 2022 Nissan LEAF that I recently purchased. This base S model had a manufacturer's suggested retail price of $27,400, but that's not the bottom line. It also came with $200 splash guards, $190 carpeted floor mats and an $80 safety kit that include both emergency and first aid items. Adding in destination charges of $1,025, the price came to $28,895 before sales tax, license and miscellaneous documentation fees we will get to later.

For just under $30,000, the LEAF S as a base vehicle is hardly a stripped down, bare bones model you might remember from your dad's day when it might have no carpeting, roll-up windows and no air conditioning.

The LEAF S is equipped with a 40-kWh battery and a 147-horsepower electric motor. You have power-assisted front and rear disc brakes, the e-Pedal mode for one-pedal driving, hill start assist and a portable charging cable that plugs into either a 120- or 240-volt outlet.

Nissan prides itself on the level of safety features that come standard on all its vehicles and the LEAF is no different. It features automatic emergency braking with pedestrian detection, rear automatic braking to keep you from bumping into things behind you, and rear cross-traffic alert. You also get automatic high-beam headlights along with lane departure and blind spot warnings with assists to keep you in your lane. In addition to a complete set of front and side airbags, standard equipment includes an anti-theft system, child car seat anchors and tire pressure monitoring.

The base LEAF also comes with power windows with one-touch up and down for the driver, a rear-view camera, rear door alert in case you've left something or someone in the back seat, and keyless locking and starting. Cruise control is standard, although you need to step up a trim level to get an adaptive system with a stop-and-go function for use in heavy traffic.

You also get heating and air conditioning with a pre-conditioning mode and an infotainment system with an 8-inch screen, Apple CarPlay/Android Auto compatibility, three months' free Sirius XM satellite radio, Bluetooth connectivity, Siri Eyes Free, voice recognition and hands-free text messaging assistant.

If you spend $1,400 more for the SV trim, add 17-inch alloy wheels instead of 16-inch steel wheels with plastic covers, the adaptive cruise control, 6 speakers rather than 4-speaker audio, navigation, heated seats and Nissan Connected capability that allows you to manage the recharging and pre-conditioning of your vehicle via your smartphone.

The Nissan LEAF Plus S, with its 62-kWh battery and 225-mile range offers pretty much the same content as the base S for $5,000 more to cover the cost of the larger battery and a more efficient and less draining heat pump for the climate controls. While you get more range, it also takes longer to recharge at 240 volts, 11 versus 7.5 hours. Quick charging the larger battery to 80 percent takes 45 minutes compared to the standard LEAF's 40 minutes. How far to do you want to go? How much do you want to spend? The answer here is obviously $5,000 for those 75 extra miles of range.

Finding the Right Trim

From the Tightwad's perspective, the vehicle and equipment you select should match the vehicle's intended use. Part of that is where you live. Here in California, the base LEAF S with its standard set of features, smaller battery with quicker recharge time and lower price of entry works for a person who has a limited commute or is looking for a second car. We also own a Hyundai Ioniq plug-in hybrid that serves as our road trip car. There are more creature comforts on that model, like heated and powered seats, that I passed on to keep the LEAF's price down.

You might want to consider a higher trim for more range and amenities.

However, if you live in a less temperate climate, you might opt for the SV version of the basic LEAF to get the heated seats. Since they run off the 12-volt system, you have some comfort and may avoid using the range-draining resistance heater. Or if you live in the snow belt, you may want to opt for the LEAF Plus for its more efficient heat pump and the bigger battery pack to compensate for any loss in range because of extremely cold weather.

Once I decided on the LEAF S, my first stop was the manufacturer site for Nissan. In my initial search, I made sure to enter my zip code. For those worried about getting bombarded by emails and phone calls, putting in your zip code simply opens the door to whatever deals may be available in your area. New vehicle incentives are often tailored specifically to geographic areas with programs that offer money to the dealers with the savings ostensibly passed along to the consumer.

This allows the manufacturer to help sell some vehicles, like all-wheel drive models in Florida, or convertibles in the Northeast in winter, that might be slow movers. The same goes for electric vehicles. As mentioned earlier, some EVs are not offered in every state. And there may be some localized programs to help a manufacturer promote electric cars in an area where they can generate some clean air credits. The bottom line here is put in your zip code and you'll be able to see the best deals that are available to you.

Also worth checking out on the website are additional incentives for recent graduates, veterans and first responders. Usually, these incentives can be added after you've made your best deal.

Providing your zip code also allows you to see what vehicles the local dealers are currently carrying in stock or units that are expected to arrive soon. Here in South Orange County, I had access to the inventories of three dealerships in Irvine, Tustin and San Juan Capistrano. The inventory showed what vehicles were available or about to come in, their colors and a facsimile of the window sticker. The base LEAF S comes in white, silver, gray, black or blue. I wanted a blue one and found it in stock at Tustin Nissan.

There were a few tools I could use to calculate how much my monthly payment would be based on the sticker price with destination. It also let me know that there was zero-percent financing available from Nissan Credit for up to 72 months. However, I had to register if I wanted to find out the total price and figure out the out-the-door cost of the LEAF.

That's where you'll start to hear from the dealer. But it also showed that tax would be about $2,500, licensing and title another $425, a $30 fee for electric vehicle registration and $85 document processing fee, which the dealer charges. Taking a bit off the top was an instant $750 California clean fuel rebate for buying an electric car. All told the price came to just over $30,000. With 25-percent down and the zero-percent financing for six years, the payments net out to about $350 per month. That's quite a bargain when compared to the average new car payment, which according to the auto finance gurus at Experian, is close to $600 per month.

Cash, Loan or Lease?

The least expensive way to buy a car is to pay cash. But you'll probably find yourself financing some portion of the purchase price. Taking out a long car loan to keep the monthly payments low isn't advisable since you may find yourself upside down, owing more on the car than it's worth. The longer the loan term, the more interest you'll pay in total. However, in my case, since it's a zero-percent rate, the longer loan works because the car will be paid off in cheaper dollars because of inflation.

If you do have to finance, another smart shopping tip is to prequalify for a loan from your bank, credit union or other online source. Once you've determined how much you want to borrow and at what rate, ask the salesperson if they can beat that deal either through the manufacturer's credit arm or other financial institutions at their disposal.

If you have a trade-in, other strategies come into play. The first is determining what your current car is worth by going to sites like Kelley Blue Book for trade-in and private sale values. While there, you can also use the Instant Cash Offer tool to receive trade or outright purchase offers from local dealers. But don't stop there — you can access other sites like CarMax, TrueCar and Edmunds to find out what they're willing to give you.

By gathering up several of these offers, you can set a floor for the value of your vehicle. When negotiating for your new car, see what a dealer's best offer is for a cash sale and then ask them what they will give you in trade for your current vehicle. If it's below your floor price, show them your offers and ask if they can do better. If not, take your best deal and sell your car elsewhere. Given the current shortage of late model used cars, buyers who have a vehicle to trade are in the driver's seat.

What About Those Tax Credits?

Besides beating high prices at the pump, electric vehicles offer the benefit of tax credits and other low or zero-emission vehicle incentives. The big one you hear about the most is the federal electric vehicle tax credit of $7,500. However, that amount is applied to your current tax year. If you buy an electric car now, you'll be able to apply the credit to your tax bill next year when your return is due.

It's real money, but unlike a traditional rebate or incentive, it's not applied immediately when you make your purchase. That means while many electric vehicles are advertised with prices reflecting the tax credit (Tesla was famous for this by including the credit and gas savings on its website pricing), remember that you must pay the full amount that you agree as the purchase price. Some state rebates may be immediately applied to the purchase price, but more often they are sent to you after the sale.

And not all electric vehicles are eligible for the tax credit. As a result of recent tax changes, only electric vehicles and plug-in hybrids built in North America are eligible for the credit. Further, half the $7,500 credit is predicated on North American sourcing of at least 40 percent of the materials used in the battery. The balance is contingent on at least 50 percent of the battery pack's North American assembly.

As a result, a wide range of imported electric vehicles are no longer eligible for this federal tax break. Other limitations established in the law caps eligibility on both the purchase price of the vehicle (cars at $55,000, trucks and SUVS at $80,000) and a buyer's income ($150,000 filing singly, $300,000 for joint filers).

While the credit originally was geared to promote EV sales, it's now more of a policy tool to encourage North American sourcing of raw materials and assembly of electric vehicles. Although major manufacturers like Hyundai/Kia have committed to building large factories here, buyers won't see vehicles from these plants until 2025 at the earliest. Consequently, tax credits may be off the table for most EVs sold in the U.S. for quite some time.

The phasing out of tax credits is another variable EV (and plug-in hybrid) buyers will have to consider when shopping for a new vehicle. Often that credit played a large factor in bringing EV prices in line with gasoline-powered rivals. A case in point is the 2022 Hyundai Kona, which was offered with gasoline or electric powertrains. The base price of a Kona SEL with a gas engine was $23,100, while an SEL electric started at $34,000. In addition to the $7,500 tax credit, local incentives, like California's $2,750 clean vehicle rebate, brought parity to final out-of-pocket costs.

Even as more plug-in hybrids and electric models come to market and are sold, the number of companies eligible for tax credits because of recent legislation is dwindling. In the case of Hyundai, the new Ioniq 5, Ioniq Plug-in Hybrid as well as plug-in versions of the Tucson and Santa Fe are no longer eligible for the credits.

Local Incentives Still Apply

In addition to the federal tax credits, there are other state and local incentives available to electric car buyers. All electric and plug-in hybrid vehicles remain eligible for these rebates. Be aware that some of these incentives have income caps on them, which grew out of protests over wealthy buyers of expensive electric vehicles getting government subsidies.

In California, the instant $750 electric vehicle rebate is just the first in a string of incentives available to the public. After the purchase, buyers can apply for an additional $2,000 clean vehicle rebate, subject to income limitations. Single filers can't exceed $135,000 gross income, while a head of household is limited to $175,000 and married filed jointly is capped at $200,000. Low-income individuals below $60,000 are eligible up to an additional

$4,500. There is an old vehicle replacement program in certain clean air districts for low-income individuals that adds an additional $4,500 in rebates.

Other states with electric vehicle incentive programs include Arkansas, Florida and New Hampshire, which offer tax credits or rebates of up to $1,000, while Pennsylvania offers $1,500. Rebates or tax cuts of up to $2,500 are offered in Colorado, Montana, Oregon and Virginia. Connecticut's incentives range from $500 to $3,000, while Delaware offers $3,500. Louisiana's rebate is 10 percent of the cost of the vehicle up to $2,500, while Maine and New York have rebates of up to $2,000. The highest state incentives are offered by Vermont and New York, which can go up to $5,000.

What About a Lease?

Leasing is another alternative to electric car ownership but is a more complicated transaction. Since the vehicle is owned by the leasing company, lessees aren't entitled to the federal tax credit. However, some of that tax savings should be passed along or reflected in the lease rate. That's a question you should ask when going over the fine print of the contract.

Another factor influencing lease rates is the residual or resale value of the car itself. Earlier generation electrics, with their limited range and shorter battery life, were worth only a fraction of their purchase price on the used vehicle market. To cover that high depreciation, monthly lease rates would have been sky high. But oftentimes, manufacturers, anxious to get electric vehicles out on the street to generate much needed clean air credits, set unrealistically low lease rates. That's why in California, it was possible to lease a Fiat 500e electric for less than $100 per month.

Sergio Marchionne, then chairman of Fiat Chrysler Automobiles, called the 500e his "compliance vehicle." It was an indication that the company was taking a big loss to generate the clean air credits needed to continue to sell high-profit trucks and SUVs without government penalties.

Over time, electric vehicle range and battery life have improved along with their resale value. With rising values come more realistic leasing rates. For those looking to try an electric vehicle for the first time, a short-term lease might be the way to go. It's especially attractive if you're planning to use the electric as an additional vehicle in your fleet for short commutes and errands

However, be aware of mileage caps. Some attractive rates may be restricted to less than 10,000 miles per year with excess mileage charges at lease end of 25 cents per mile. Also, a low monthly lease rate may also require a large down payment or trade. Take that value, divide it by the term of the lease and add that to your estimated monthly payment to get a truer picture of your out-of-pocket expense. Also check to see if there are additional state incentives for leases. Colorado, for instance, offers a $1,500 rebate on electric car leases.

Buying Strategies

In making the decision on whether you want to buy or lease an electric vehicle, the right Tightwad strategy depends on your plans for the car. On one hand, if you're buying the cheapest electric vehicle out there and are looking to literally drive it into the ground or pass it along to a family member, then get a base model, pay cash or find zero-percent financing.

Because electric vehicles require little maintenance, there are no oil changes or coolant flushes to worry about ever, so it's an ideal candidate to become a permanent part of the family. On the other hand, if you budget your transportation needs as a monthly amount, a lease might be a good way to familiarize yourself with the EV experience. And, for the same monthly payment as a loan, you can get a higher trim or longer-range model.

Since leasing is really a long-term rental, you're only paying for the part of the car's depreciation that you're using. As a result, your total costs are less than if you purchased the vehicle outright.

At the end of the lease, if you like the vehicle, there's a buyout option. If it that price is less than what the vehicle is worth on the market, you come out ahead by buying it. If it's worth less, simply walk away.

Leasing is also an attractive option since EV technology is rapidly changing. If you're not satisfied with the range, performance or features on your current car, like the weather in Chicago, if you wait a minute, it will change. Three or four years down the line, there could come a battery technology breakthrough that greatly extends range, allows for faster recharging and brings lower costs.

 Leasing affords you the opportunity to remain flexible in your future choices and may, in fact, be worth the extra cost over outright ownership.

While electric vehicles remain more expensive than traditional gas-powered cars, there are affordable options available today. All it takes is time, patience and planning to find the deal that makes the most sense for you and your wallet.

Tightwad Pro Tip:

When shopping for an affordable electric car, check to see if the vehicle comes with a portable 120/240-volt EVSE (Electric Vehicle Service Equipment) charging cord. Increasingly, manufacturers are eliminating this accessory or charging extra for it. Buying one may add another $200 to your purchase and you may need it in any event if you decide to pass on installing a more expensive Level 2 charger in your garage.

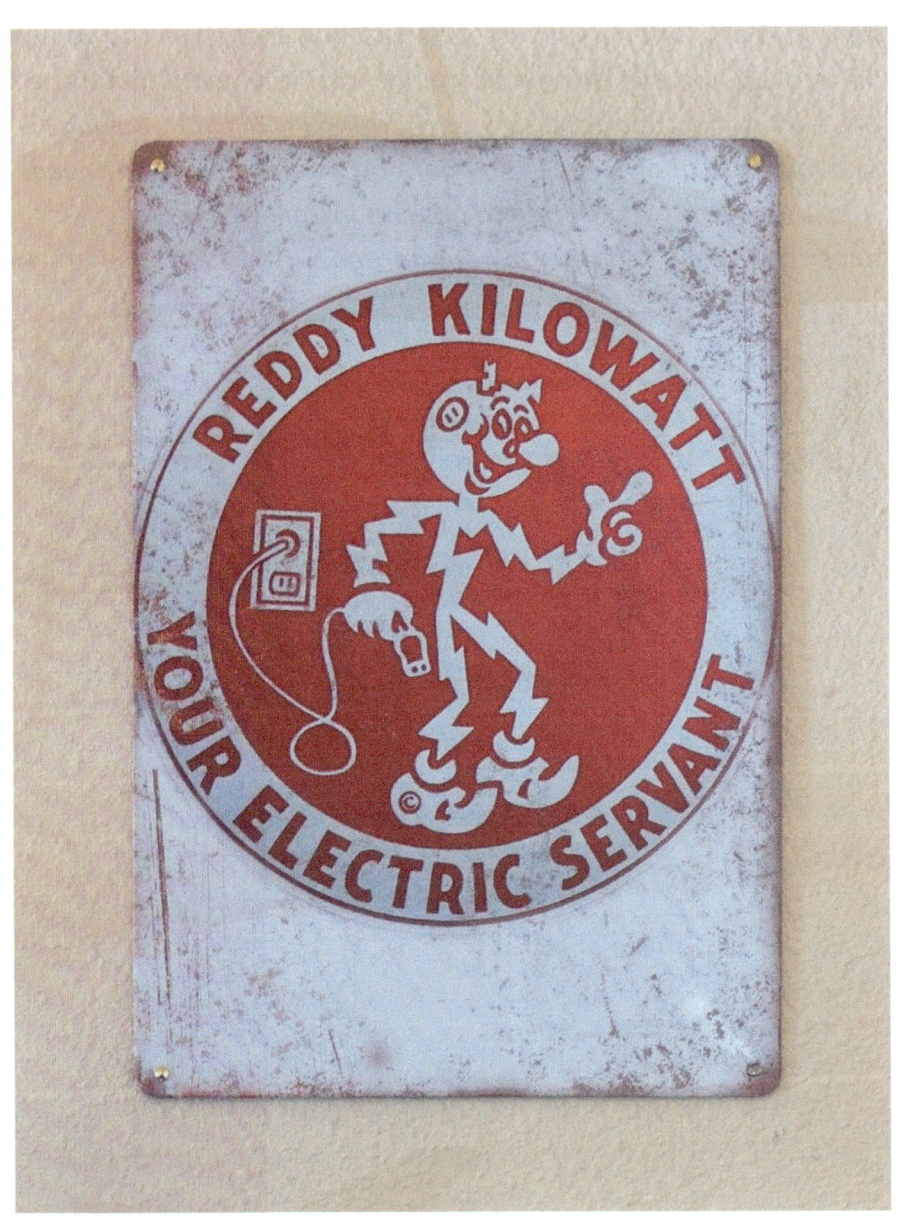

6. THE TIGHTWAD'S GUIDE TO ELECTRIC CARS

Now that we've learned how electric cars work, what they're like to drive and how to shop for one, it's time to get down to brass tacks. What's out there? Currently, there are eight electric vehicles for sale with manufacturer suggested retail prices (MSRP) below $40,000 and another nine with base prices between $40-50,000. Above $50,000, there's more than a dozen electrics ranging from the Audi Q4 e-tron up through the Lucid Air that tops out at nearly $170,000.

The list here, arranged from lowest to highest prices, is divided as Tightwad Affordable and Tightwad Stretch with $40,000 as the dividing line. The MSRPs reflect available 2022 and early 2023 model year pricing.

TIGHTWAD AFFORDABLE

Nissan LEAF

The Nissan LEAF starts at $28,040 for the 40-kWh battery version offering 149-mile EPA range. It's a five-passenger hatchback with front-drive. All models are capable of both 120/240 AC and DC quick charging with dual receptacles. The LEAF SV Plus, with its 62-kWh battery and range of 225 miles, starts at $36,040.

It comes in just the two trims, S and SV Plus and features a freshened exterior for the 2023 model year that includes a new front fascia and a different wheel design from the 2022 model. The SV trim level from the base version has been dropped.

Nissan says the standard LEAF can be fully charged at Level 2 in 7.5 hours and to 80 percent in 40 minutes using DC fast charging. The LEAF Plus takes 11 hours at Level 2 and 45 minutes for an 80 percent fast charge.

The standard bumper-to-bumper warranty is 3 years/36,000

miles, while the electric motor is covered by a 6-year/60,000-mile powertrain warranty. The battery pack carries an 8-year/100,000-mile warranty that "covers any repairs needed to return battery capacity to a level of nine remaining segments on the vehicle's battery capacity level gauge." That reading on the 12-segment scale works out to about 75 percent capacity.

The Nissan LEAF is sold in all states and because it's built in the North America, it remains eligible for the $7,500 federal electric vehicle tax credit.

MINI Cooper SE

The electric MINI Cooper SE starts at $29,900; its price remains unchanged from the 2022 to the 2023 model year. The Cooper SE comes in a two-door hardtop version that seats four in a small hatchback package.

The 181-horsepower electric motor drives the front wheels and puts out 199 lb-ft of torque which MINI says is good for a 0-60 mph time of 6.9 seconds. MINI is renowned for its go-kart-like handling and the Cooper SE is no exception.

For 2023, upgrades include standard Apple CarPlay, a larger 8.8-inch infotainment touchscreen, and standard lane departure warning. If you're looking to go all in, the top MINI Cooper SE model is the Resolute package, which retails for $36,900. The package includes 18-inch black wheels, special graphics and accents and a cloth/leatherette upholstery.

The MINI Cooper SE is equipped with a rather small 32.6 kWh battery pack that has a range of 114 miles. MINI says it takes five hours to recharge the battery with Level 2 Charging, and with DC fast charging, 80-percent of battery capacity is reached in 36 minutes.

The Cooper SE comes with a 120-volt charging cord, which the company says replenishes the battery at a rate of four miles per hour, which would be over 25 hours given the 114-mile range of the pack. A home Level 2 charger is at the owner's expense and the Cooper SE has the ability for DC fast charging at public kiosks.

MINI offers a standard 4-year/50,000-mile bumper to bumper warranty and an 8-year/100,000-mile battery warranty that will replace the unit if it falls below 70-percent efficiency. The MINI Cooper SE is available in all states and is no longer eligible for the $7,500 federal electric vehicle tax credit.

Chevrolet Bolt/Bolt EUV

The Chevrolet Bolt comes in two distinct models: the hatchback, which rides on a 102.4-inch wheelbase, and the slightly larger Bolt EUV, sort of a crossover SUV with a 105.3-inch wheelbase. That extra length translates to more rear seat legroom but also a touch less cargo space than the hatchback, 16.3 versus 16.6 cubic feet.

Pricing for the 2022 Bolt starts at $31,500 for the hatchback and $33,500 for the EUV. For 2023, GM is dropping prices to a respective $25,600 and $27,200 to remain competitive.

Both vehicles use identical 200 horsepower electric motors with 266 lb-ft of torque to drive the front wheels. The Bolt employs a liquid-cooled 65-kWh lithium-ion battery pack good for 259 miles in the hatchback and 247 miles in the EUV. The difference in range is attributed in part to the EUV's 3,680-pound curb weight which is about 100 pounds more than the hatchback.

There's also a slight difference in the time it takes to recharge both. At 120 volts, both get 4 miles of range per hour, which means it can take as long at 60 hours to recharge. Chevrolet says the hatchback can be fully charged in about 7 hours at 240-volt Level 2 charging and obtain about 100 miles of range in 30 minutes using DC fast charging (that represents less than a 50 percent charge – getting to 80 percent can take closer to an hour). The EUV specs indicate that at 240-volt Level 2 charging at 32 amps, battery replenishment will take 10 hours. A 48-amp setup takes 7 hours. A DC fast charge gets 95 miles in 30 minutes.

With both, buyers get a dual-level charge cord capable of both 120- and 240-volt charging and standard DC charging capability. All models boast wireless Apple CarPlay and Android Auto connectivity, a new 10.2-inch infotainment touchscreen, enhanced safety assists and a choice of two trim levels, 1LT and 2LT on the hatchback and LT and Premier on the EUV. The latter comes with adaptive cruise control as standard equipment. The Bolt also boasts One Pedal Driving and regenerative braking on demand via a paddle mounted to the left side the steering wheel.

The standard warranty is a 3-year/36,000-mile bumper-to-bumper coverage with an 8-year/100,000-mile battery warranty including replacement if the battery falls below 60 percent capacity.

The Chevrolet Bolt is sold in all states and regained eligibility for the $7,500 federal electric vehicle tax credit on Jan. 1, 2023.

Chevrolet Equinox EV

Chevrolet is promising its second affordable EV in the form of the

Equinox EV, a compact SUV with a projected base price of $30,000. It's slated to bow in the fall of 2023 as a 2024 model.

It will be available in a variety of trims, the base 1LT will come in front drive with a range of 250 miles. An extended range version will offer a 300-mile range, while an all-wheel drive variant with the same larger battery pack will go 280 miles between charges.

Front-drive models will produce 210 horsepower and 242 lb-ft of torque, while the all-wheel drive versions are rated at 290 horsepower and 346 lb-ft of torque. Expect the two Ultium battery packs to offer somewhere in the range of 60-90 kWh.

With standard 11.5 kW Level 2 charging, the 250-mile version can be recharged in about 7.5 hours, while the 300-mile battery will take about nine hours. At 19.2 kWh, the top 3RS model delivers 51 miles of range per hour.

Unfortunately, the first models offer will be a limited edition 3RS version in long-range form, which will likely cost considerably more than the bandied $30,000 entry level price tag. The least expensive variants may not be available until well into 2024.

The standard warranty is a 3-year/36,000-mile bumper-to-bumper coverage with an 8-year/100,000-mile battery warranty including replacement if the battery falls below 60 percent capacity.

The Equinox EV will be available in all 50 states and will be eligible for the $7,500 federal tax credit.

Mazda MX-30

Available only in California, the Mazda MX-30 is the Japanese brand's first effort at electrification. In fact, Mazda is one of the few makers that has never offered a hybrid, much less a battery

electric vehicle. The new compact SUV is priced from $33,470.

It has a small 35.5 kWh battery pack good for a range of only 100 miles. The 145-horsepower electric motor drives the front axle, and 0-60 mph is said to take a leisurely 9.6 seconds. The battery takes 21 hours to recharge on 120 volts and about six hours with Level 2 240-volt service.

The vehicle comes only with a Level 1 120V charge cord. Buyers get a $500 ChargePoint credit towards a home charger or use of the company's public charging stations. Mazda says it takes 36 minutes for an 80 percent charge using a DC fast charger.

Priced higher and offering the shortest range of affordable electrics, Mazda put a premium on the look and interior fitments of the car. Also, Mazda prides itself on spirited handling and early reports are favorable about the vehicle's dynamics. It is a handsome design with some unusual features like the rear-hinged back doors.

Among the high-end amenities on the car are a heat pump for the climate controls, adaptive cruise control with stop-and-go capability, and power moonroof. Mazda also offers MX-30 buyers free use of non-electric loaners for up to 10 days per year.

The Mazda MX-30 comes with a standard 3-year/36,000-mile bumper-to-bumper warranty and an 8-year/100,000-mile battery warranty which will replace batteries that fall below 70 percent efficiency. The Mazda MX-30 will be initially sold in California only and is not eligible for the $7,500 federal electric vehicle tax credit.

Hyundai Kona Electric

This small SUV is the electric variant of the popular Kona gas-powered model. Refreshed for 2022, the Kona Electric is distinguished from the gas variant by having a solid front fascia instead of a traditional grille. Pricing starts at about $34,000.

Power comes from a liquid-cooled 64-kWh battery pack that drives the front wheels. The 201-horsepower electric motor develops 291 lb-ft of torque. Hyundai says the Kona Electric is good for a range of 258 miles, one mile less than its direct competitor, the Chevrolet Bolt Hatchback.

Buyers will find that the Kona comes only with a 120-volt charging cord; installing Level 2 equipment is at the owner's expense. Using a standard outlet, it takes about 50 hours to recharge the Kona's battery pack. With a Level 2 setup, that time drops to about 9.5 hours, and at 100-kW fast charging, 80-percent capacity is available in 47 minutes.

Kona Electric comes in three trim levels, SE, SEL and Limited. The SEL is priced from $37,300 while Limited starts at $42,000. Buyers in cold climates may opt for the Convenience Package on the SEL which includes a battery warmer system.

Also note that earlier Kona models had less-efficient electric resistance heating instead of the current heat pump. The Convenience package, which adds $3,500 to the price, includes a sunroof, power driver's seat, wireless device charging and heated front seats.

All models come with a 10.25-inch infotainment screen with wireless Apple CarPlay and Android Auto compatibility. The Limited costs considerably more at $42,500 and includes the features in the Convenience package plus adaptive cruise control with stop-and-go, ventilated front seats, premium audio, and leather upholstery. Unlike the SEL, the Apple CarPlay and Android Auto are not wireless.

The Kona Electric comes with a 5-year/60,000-mile basic bumper-to-bumper warranty. The battery carries 10-year/100,000-mile coverage against failure or degradation below 70 percent. The Kona is sold in all states and is not eligible for the $7,500 federal electric vehicle tax credit.

Kia Niro Electric

While Kia pitches the Niro Electric as an SUV, this front-drive hatchback looks more like a wagon than an off-roader. Niro is

offered in hybrid (with standard and plug-in versions) and full battery electric. The Niro slips in under the Tightwad Affordable wire at $39,950.

Beneath the skin, the Niro shares its electric drivetrain with its corporate cousin, the Hyundai Kona. Power comes from a liquid-cooled 64-kWh battery pack and drives a 201-horsepower electric motor, which also makes 291 lb-ft of torque. Unlike the 258-mile Kona, Niro's range is rated at 253 miles, part of which can be attributed to its slightly heavier curb weight.

Using the standard Level 1 Charger, Kia says it will take 2.5 days to fully recharge the battery pack. Level 2 charging is a much more reasonable 9 hours and 35 minutes, while 100 kW fast charging will give the Niro EV an 80-percent level in an hour.

The Kia Niro comes in two trim levels, Wind and Wave. A battery warmer system and heat pump are listed as options. The Niro rides on 17-inch alloy wheels and includes standard adaptive cruise control with stop-and-go features, heated front seats, power adjusted driver's seat, 10.25-inch infotainment touchscreen, wired Apple CarPlay and Android Auto.

An advantage of the Niro's wagon-like profile is the 53 cubic feet of cargo space available when the rear seat is folded down.

The Kia Niro Electric comes with a 5-year/60,000-mile bumper-to-bumper warranty and a 10-year/100,000-mile battery warranty. The 2023 Niro EV is available in all 50 states and is not eligible for the full $7,500 federal electric vehicle tax credit.

TIGHTWAD STRETCH

While the average new car transaction price hovers around $45,000, the Tightwad knows that's pushing the outer limits of affordability when it comes to buying new. Still, with the federal tax credit and other incentives, that doesn't necessarily mean that an electric car with a sticker price between $40,000 and $50,000 is beyond the realm of possibility.

There are a number EVs on or about to come to the market in this space. Be forewarned though, while the base price plus tax credit may push the out-of-pocket expenses below $40,000, many of these models are entry level trims. As such, they're great fodder for ads and other come-ons but finding one and buying it at the advertised price is another matter. Caveat emptor.

Volkswagen ID.4

The Volkswagen ID.4 is a compact SUV that replaces the more affordable eGolf in the company's lineup. While it carries a sub-$40,000 sticker of $38,995, its price for most models in the lineup with destination will be over that mark. VW boasts that the ID.4 with optional front and rear electric motors is the least expensive all-wheel drive EV with the list price of $47,995.

The ID.4 comes in a number of trim levels including standard, S,

Pro, Pro AWD, Pro S and Pro S AWD. The standard and S models feature an 62-kWh battery pack, while the Pro models use an 82 kWh battery pack. A 201-horsepower electric motor with 229 lb-ft of torque drives the rear axle, while the all-wheel drive version adds a front 107 horsepower motor with 119 lb-ft of torque that contributes to a combined system output of 295 horsepower.

The base under $40k standard model as well as the $43,995 ID.S have a range of only 209 miles. Rear-drive Pro models can go 275 miles on a single charge, while all-wheel drive versions are rated at 255 miles. Level 2 charging takes 7.5 hours, while DC fast charging of 125-135 kW can bring an 80 percent charge in about 30 minutes. VW, which helped establish Electrify America charging network, gives buyers three years of unlimited 30-minute sessions at no charge.

Among the standard features on the ID.4 are 19-inch alloy wheels, adjustable regenerative braking through a Brake (B) mode on the drive selector and a standard 10-inch infotainment touchscreen with Apple CarPlay and Android Auto connectivity.

The Volkswagen ID.4 is covered by a 4-year/50,000-mile general warranty and 8-year/100,000-mile battery coverage with replacement if capacity falls beneath 70 percent. The Volkswagen ID.4 is sold in all states and is still eligible for the $7,500 federal electric vehicle tax credit.

Kia EV6

Like the Volkswagen ID.4, the Kia EV6 is a compact SUV electric that offers both front- and all-wheel drive versions. The base rear-drive 2022 EV6 Light which started at $40,900 has been dropped from the line. The Wind and GT-Line are $48,700 and $52,900 respectively for rear drive. Adding AWD pushes the Wind price up $3,900 and the GT-Line by $4,700. A range topping GT AWD costs $61,600.

Wind and GT-Line have 77.4-kWh battery packs with 225-horsepower motor driving the rear wheels. These models, which also make 258 lb-ft of torque, are good for 310 miles between charges.

The Dual Motor e-AWD Wind and GT-Line incorporate a front 74 kW (99-horsepower) unit that bumps total system output to 320 horses and 446 lb-ft of torque. Both models boast a 0-60 time of

4.6 seconds and respective ranges of 282 and 262 miles. The GT version offers 576 horsepower and 545 lb-ft of torque from its AWD system but has a range of only 209 miles.

The EV6 comes standard with a 120-volt outlet that can be used to draw power from the battery pack for up to 36 hours. EV6 can also use that capability to recharge another electric vehicle. Level 1 replenishment takes 68 hours for the Wind and GT-Line. Level 2 charging takes a respective 5 hours 50 minutes and 7 hours 10 minutes. Tightwads take note: the EV6 does not come with a portable 120/240V ESVE charging cord setup, so you're on your own if you want to do Level 1 or 2 charging at home.

The battery pack can accept up to 350 kW DC fast charge at 800 volts providing 80 percent capacity on both battery packs in 18 minutes. DC fast charging with a 50kW system takes 63 minutes for the Light and 73 minutes for the Wind and GT-Line.

Standard equipment on the Light includes such upscale touches as adaptive cruise control, a 12.3-inch infotainment touchscreen with navigation, Apple CarPlay and Android Auto connectivity. The seats are a combination of cloth and leatherette, and you get a full suite of driver and safety assists including blind spot warnings, lane keeping warning and assist, and heated front seats.

The Kia EV6 comes with a 5-year/60,000-mile general warranty and a 10-year/100,000-mile battery warranty. It's sold in all states but doesn't qualify for the $7,500 federal EV tax credit.

Toyota bZ4X

While Toyota briefly offered a pure EV in a California-only RAV4 built with Tesla from 2012-14, the 2023 bZ4X is its first dedicated electric aimed at the mass market. This compact SUV is about the

size of the RAV4 and is priced from $42,000.

In base trim, the bZ4X goes 252 miles on a charge from its 71.4-kWh battery pack. The more feature-laden $46,700 Limited has a range of 242 miles. Output is delivered through a 201-horsepower electric motor mounted on the front axle that produces 196 lb-ft of torque. Dual motor versions employ smaller 107-horsepower motors front and rear that produce a total system output of 214 horsepower and 248 lb-ft of torque.

All-wheel drive models, which start at $44,080, employ a slightly larger 72.8-kWh battery delivering a range of 228 miles. The AWD Limited, which costs $48,780, has a shorter range of 222 miles.

The bZ4X is capable of Level 1, 2 and DC fast charging. The 6.6-kW onboard architecture allows for Level 2 charge in about nine hours. Toyota is offering the option of including the price of a ChargePoint Level 2 home charger in the purchase price. Buyers also get a year's worth of unlimited charging at EVgo kiosks.

Among the features buyers can expect are a 12.3-inch infotainment touchscreen with wireless Apple Car Play and

Android Auto connectivity and climate controls that include a heat pump The Toyota Safety Sense 3.0 suite features automatic emergency braking, blind spot monitor and a safe exit assist that warns of approaching vehicles or cyclists. In addition to optional heated/ventilated seats, the bZ4X offers an optional radiant heat system in the floor for your legs and feet.

Toyota's basic bumper-to-bumper warranty is 3-year/36,000-mile coverage, while the powertrain has a 6-year/60,000-mile term. The battery has an 8-year/100,00-mile guarantee. The Toyota bZ4X will be sold in all 50 states, but doesn't qualify for the federal tax credit.

Hyundai Ioniq 5

Just as the Hyundai Kona Electric shares its underpinnings with the Kia Niro EV, so does the Hyundai Ioniq 5 employ the same platform as the Kia EV6. Hyundai lists a base Ioniq 5 SE Standard Range model with a 58-kWh battery and a range of 220 miles for $41,450.

The step-up model is an SE trim level featuring a 77.4-kWh battery and a range of 303 miles. It retails for $45,500. Rear-drive SEL

models list at $47,450 with the top Limited model going for
$52,600. Adding the second motor for a total output of 320
horsepower ups the price about $4,000 on all trims. It also
reduces the range to 256 miles. The Ioniq 5 does not come with a
portable 120/240V EVSE cord.

Hyundai says the larger battery takes 6 hours and 43 minutes to
charge at Level 2, standard DC charging brings 80 percent in about
25 minutes, while 800-volt DC charging does the trick in about 18
minutes. The battery packs are liquid cooled and all-wheel drive
models feature heat pumps as standard equipment.

The Hyundai Ioniq 5 comes with a 5-year/60,000-mile general
warranty and a 10-year/100,000-mile battery coverage that will
replace it if capacity falls below 70 percent. The Ioniq 5 will be
rolled out in 16 states (California, Connecticut, Maine, Maryland,
Massachusetts, New York, New Jersey, Oregon, Rhode Island,
Vermont, Texas, Florida, Illinois, Georgia, North Carolina,
Pennsylvania and Arizona) before going national. It no longer
qualifies for the $7,500 federal electric vehicle tax credit.

Ford Mustang Mach-E

Pony Car fans may have been skeptical, but Ford has successfully
grafted the Mustang legacy on an electric SUV. The Ford Mustang
Mach-E is a five-passenger compact crossover offered with the
choice of rear- or all-wheel drive. Starting at a base MSRP of just
under $47,000, the window sticker quickly gallops away when you
move up the trim ladder.The least expensive model is the
standard range Select, which delivers 247 miles with rear-drive
and 224 miles when equipped with all-wheel drive using a 70-kWh
battery. The Premium, which starts at $55,000 offers the same
performance. All-wheel drive is an extra $2,700. An extended

range version of the Premium using the 91-kWh battery pack adds $8,300 respectively to rear- and all-wheel drive versions. Range increases to 303 miles in the former and 277 for the latter.

The top GT, which boasts 480 horsepower and 600 lb-ft of torque, starts at $70,000 and goes 270 miles on a full charge in rear-drive, while the AWD version gets 10 miles less.

Naturally, with the Mustang branding, Ford touts performance. The GT is capable of 0-60 mph acceleration of 3.5 seconds, while the extended-range Premium model clocks in at 6.1 seconds. The Select, thanks to its less dense and lighter battery pack, turns the trick in 5.8 seconds.

Using Level 2 charging, the EPA says the standard range Mustang Mach-E takes 8 hours to recharge, while the extended range versions take nearly 11 hours. Ford equips all models with a charge cord for both 120- and 240-volt outlets. Due to high demand, is only taking orders for the 2023 model year.

The Ford Mustang Mach-E comes with a 3-year/36,000-mile

bumper-to-bumper warranty with 8-year/100,000-mile coverage for the battery pack. The Mach-E is available in all states and is eligible on for the $7,500 federal electric vehicle tax credit.

Chevrolet Blazer EV

Slated to launch in summer of 2023, the Chevrolet Blazer EV is GM's answer to the Mustang Mach-E. This sporty electric five-passenger SUV will be available in front-, rear- and all-wheel drive configurations depending on trim level. These range from a base 1LT up through more performance-oriented RS and SS models.

The base 1LT model, which will be front-drive only, will offer a range of 247 miles and retail for $44,995. However, it won't be available until early 2024. The 2LT, which will also be front-drive with an AWD option, will have a range of 293 miles and will start at $47,595.

The sportier RS, which stickers at $51,995, boasts a range of 320 miles and can be specified in front-, rear- or all-wheel drive. The top all-wheel drive SS model, which will retail for $65,995, is rated at 557 horsepower and 648 lb-ft of torque. It will have an expected range of 290 miles between charges. The SS won't be available until late 2023.

Among the standard features are a 17.7-inch infotainment touchscreen and 11-inch digital instrument cluster. Chevrolet says the Blazer will also have the capability of 11.5 kW Level 2 charging and standard DC public fast-charging capability of up to 190 kW, depending on the model. The latter will allow about 78 miles of range to be added in 10 minutes, per GM estimates.

The standard warranty is a 3-year/36,000-mile bumper-to-bumper coverage with an 8-year/100,000-mile battery warranty including replacement if the battery falls below 60 percent capacity.

The Blazer EV will be available in all 50 states and will be eligible for the $7,500 federal tax credit.

Subaru Solterra

The first all-electric vehicle from Subaru, the Solterra, is a twin of the Toyota bZ4X. The program is the opposite of the joint venture between the two automakers that has Subaru producing both the BRZ and Toyota GR 86 sport coupes. This time around, it's Toyota that makes both vehicles.

Keeping to its off-road heritage, Subaru will offer Solterra only in a higher trim with dual-motor all-wheel drive. The specs – 218 horsepower output and 248 lb-ft of torque with a range of about 220 miles – are comparable to that of the AWD bZ4X. Solterra uses the slightly larger 72.8-kWh battery pack. Subaru says that DC fast charging will give 80 percent capacity in under an hour.

Subaru also is stressing the vehicle's off-road capability by highlighting its 8.3 inches of ground clearance, which aligns with its gas-powered all-wheel drive compact SUVs.

Like its Toyota sibling, the Subaru is equipped with a 12.-3-inch infotainment touchscreen with wireless Apple CarPlay and Android Auto compatibility. Pricing starts at $44,995 and tops out at $51,995 for the top Touring trim.

The Subaru Solterra will be sold nationally and comes with a standard 3-year/36,000-mile warranty as well as 8-year/100,000-mile battery coverage. The vehicle is not eligible for the full $7,500 federal electric vehicle tax credit.

Tesla Model 3

Originally, the Tesla Model 3 midsize sedan was supposed to be the company's $30,000 entry-level vehicle. Even as it's ramped up to be the second most popular in the lineup behind the Model Y crossover, its pricing remains above $40,000.

The least expensive model you can buy is a rear-wheel drive standard range model at $43,990. It comes with a 54-kWh battery pack and can go up to 267 miles on a charge. Stepping up to a dual-motor, all-wheel drive Performance model adds $10,000 to the price. That version has a range of 315 miles.

The standard range model comes with a 228-horsepower electric motor. The Performance models feature two motors with a combined output of 335 horsepower. At 240-volt Level 2, the EPA says it will take 10.4 hours to recharge the battery and only 5.8 hours using a 48-amp outlet. The larger battery takes a respective 11.5 and 9.6 hours.

The Tesla Model 3 comes with a 4-year/50,000-mile general warranty and battery coverage of 8 years/120,000 miles with replacement of batteries that fall below 70 percent capacity. The Tesla Model 3 is sold online only. While Tesla maintains showrooms and service facilities in most states, due to franchise dealer laws in Connecticut, Michigan, Louisiana, Texas, Utah and West Virginia, vehicles must be delivered from out of state. The Tesla Model 3 is again eligible for the $7,500 federal electric vehicle tax credit.

Polestar 2

Polestar is a new electric car company that has grown out of what formerly was a performance car sub-brand at Volvo. It launched its first vehicle, a plug-in hybrid called Polestar 1, before switching over to what will become a lineup of purely electric vehicles in the

Polestar 2 five-door hatchback and Polestar 3 compact SUV.

The least expensive Polestar 2 on the market is a front-drive, single motor, four-door liftback sedan that employs a 78-kWh battery pack good for a range of 270 miles. For 2023, it starts a $48,400. A 231-horsepower electric motor producing 243 lb-ft of torque enables the car to accelerate to 60 in 7.0 seconds.

A dual motor version with all-wheel drive delivers 249 miles per charge from the same size battery pack. The extra motor bumps total system output to 408 horsepower and 487 lb-ft of torque for a 0-60 mph time of 4.5 seconds. The 2023 dual motor version starts at $51,900. It's surprising at this price point a heat pump for the climate control is an extra cost option.

The Polestar 2 comes with a charging cable for 120- and 240-volt charging. At the lower voltage, it will take about 72 hours to recharge the battery, while Level 2 charging takes about 10 hours. The Polestar 2 is one of the first cars on the market to use Google's operating system for the infotainment and, as a result, you get Google Maps and Android Auto compatibility, but no Apple CarPlay.

The basic warranty on the Polestar 2 is 4 years/50,000 miles, while the battery is covered for 8 years/100,000 miles. Polestar 2 is sold online, although there are some showrooms and service centers at 23 locations along the East and West Coasts. The brand also has locations in Texas, Georgia, Michigan, and Minnesota. The built-in-China Polestar 2 is not eligible for the $7,500 federal electric vehicle tax credit.

Nissan Ariya

Nissan has a new compact SUV called Ariya priced from $43,190. Available in eight trim levels and two battery configurations, Ariya offers front- and all-wheel drive starting with the entry level Engage. Step-up trims include Venture +, Evolve+ and Platinum+ trims, a flagship with all-wheel drive tops the lineup at $60,190.

Engage models with 63 kWh battery packs employ a 238-horsepower electric motor making 221 lb-ft of torque. Range is

216 miles. With the larger 82 kWh battery pack, the lighter weight Venture+ is good for a 300-mile range, while the more feature-laden Evolve+ and Premiere+ will go 285 miles between recharging. The Platinum+ will boast total output of 389 horsepower and 442 lb-ft of torque from its all-wheel drive setup but will only go 265 miles on a charge.

With its debut as a 2023 model, the Ariya is covered by Nissan's 3-year/36,000-mile general warranty and 8-year/100,000-mile battery coverage. It will be sold in all states and but since it is built in Japan, it is not eligible for the $7,500 federal tax credit.

NOW A WORD ABOUT ELECTRIC PICKUPS

Ford F-150 Lightning

There's a lot of noise about electric full-size pickups, most notably the Ford F-150 Lightning, Chevrolet Silverado EV and the Tesla Cybertruck. At first blush, the base prices being bandied about for these pickups, from as low as $39,900 for the Tesla and Silverado to $41,995 for the Lightning, seem affordable.

But looking closer at the specs, you'll see that these vehicles come with smaller battery packs and limited range. Recent price increases on the Ford have bumped the entry level price to nearly $56,000. In addition, these "work trucks" won't offer the same level of amenities that car or SUV shoppers expect when buying a vehicle for personal use. Step out of the work truck category to an XLT trim of the F-150 Lightning and you'll pay $63,474. The first Chevrolet Silverado EVs to go on the market in late 2023 will be the RST First Edition model priced at $105,000. The top Lightning Platinum model will retail for $97,000. While Tesla initially announced a dual-motor Cybertruck starting at $49,900 and tri-motor version at $69,900, it has since removed the pricing and spec info for both from its website.

Then there's the tax credits, or lack thereof. As a result of recent legislation, pickups and SUVs are capped at $85,000 for tax credit eligibility. Plus, there are new income rules that limit the credits to individuals with gross income of less than $150,000 and joint filers below $300,000. Electric SUVs not built in the U.S. are also no longer eligible to receive the tax credit.

The other big question is affordability vs. capability. Part of the allure of driving a pickup, other than the cool factor, is the ability to either haul cargo or tow things. This additional work affects range and it's a difficult question to answer because the weight involved has high variability. Manufacturers have yet to produce even ballpark estimates of the impact towing or hauling will have on the range of these vehicles. Since the cheapest versions of these trucks will also have the smallest batteries and shortest range, buying a pickup to do what pickups do may defeat the purpose of going electric in the first place.

7. IS BUYING A USED ELECTRIC CAR A GOOD IDEA?

Solid Tightwad advice is that if you can't afford a new car, look for a low-mileage, late model used one instead. It's been a good strategy, at least until the recent shortage of new cars has greatly boosted used car prices. During the beginning of 2022, Kelley Blue Book reported that an average used car was selling for over $28,000. That price represents a 28 percent year-over-year gain and a whopping 42 percent increase from 2019.

Still, with the average new vehicle transacting at $45,000, according to KBB, a used vehicles still represents a savings over buying a new car. Does the same logic apply to electric cars? Well, yes and no. It depends on the age of the vehicle and the type of battery pack and range it offers.

All the other considerations we've gone over regarding a new purchase apply here as well. These questions include how it's going to be used, if there are other cars available for spur of the moment trips outside the electric's normal operating range, and access to charging.

The cheapest used electrics are typically those first-generation models that, when new, could go maybe 90 or at best 100 miles on a charge. Some of these vehicles have battery degradation issues and may now only offer 50 or 60 miles on a charge.

A used Fiat 500e is a bargain but may not meet all your needs.

Also keep in mind that many of these earlier vehicles won't have quick charge capability. Later and more improved models with 125- to 150-mile range typically may cost upwards of $5,000 to $10,000 more than the cheapest used EVs on the market.

Once you get into later model vehicles, especially those with more than a 200-mile range, you'll be paying near new prices. For 2017 and later models, which offered a 125-mile range and quick-charge capability the earlier models didn't have, you'll find prices upwards of $20,000. With the rebates and tax credit, a new 2022 Nissan LEAF was cheaper, could go farther on a charge and, unlike a used model, had only five miles on the odometer.

However, a recent change included in the Inflation Reduction Act provides a tax credit of up to $4,000 on used electric vehicles two years old or older that cost no more than $25,000 purchased from a licensed dealer. There are a few catches. In addition to the age and cost limitations, eligibility is extended only to those single tax filers making less than $75,000, up to $150,000 for joint filers or below $112,500 for head of households.

The credit can be applied only once a particular vehicle and is pro-rated at 30 percent of the vehicle's sale price (i.e. $3,000 on $10,000 car) up to the $4,000 maximum. Like all these tax credits, the buyer must pay the full amount of the transaction and apply for the credit when they next file their taxes. However, starting Jan. 1, 2024, the new law allows the used vehicle tax credit to be applied at the point of sale.

There are also local incentives to buy a used electric car. In California, there are programs by local air quality districts that offer up to $4,500 in rebates based on income to purchase used electrics. Local utilities also have incentives ranging from $1,000 to $1,500.

That's not to say there aren't incentives to buy a used electric car. In California, there are programs by local air quality districts that offer up to $4,500 in rebates based on income to purchase used electrics. Some local utilities have incentives ranging from $1,000 to $1,500. You may also get money for scrapping older gas cars when you by a new or used EV. But be sure that amount you get for your clunker is more than the vehicle's trade-in value before deciding to scrap it.

What to Look for in a Used Electric Car

When shopping for a used electric vehicle, you need to consider many of the same things that you would on any ordinary pre-owned car. Most important is the number of miles and the age of the vehicle. Both have an impact on its remaining life. You will also want to get a vehicle history report to determine if it's been in a crash or if there have been any major repairs.

Usually wear items, like tires, windshield wipers and brakes, should be checked first. Thanks to regenerative braking, electric vehicles typically will see much longer service life on things like brake pads and rotors.

You should also take the vehicle for a test drive to determine if the alignment is good and that the shocks are firm. Also, it will help you to determine whether there are any squeaks or rattles that need to be addressed and how well the accessories like heating and air conditioning function.

You should also ask to see any maintenance records. This is vital with a conventional internal combustion engine car in determining if it's been regularly maintained with oil changes, coolant flushes and other regular service items. While EVs don't have those needs, there are things that do need to be attended to regularly on an electric, including cabin air filter replacement, tire rotation, wheel alignment and shocks.

These records would also show if there were any unusual service events, including the replacement of the 12-volt electric battery, failed electronics or major battery issues. You should also check the National Highway Traffic Safety Administration website for any recalls.

Find out if the vehicle was garaged or not. Exposure to temperature extremes, especially cold without a battery warmer, can affect battery life. Also, if the vehicle has quick charging capability, it's best to ask how often it was used. Relying primarily on DC fast charging over Level 1 or 2 AC charging may shorten the life of the battery.

Check that Battery Pack!

The biggest concern in buying a used electric car is the health of the battery pack. It's the most expensive component and you can't just take it to your corner mechanic to get it checked out. Virtually all electric vehicles are equipped with on-board diagnostics and readouts that will give you a general idea of the remaining capacity on the battery.

Getting a full check on the health of the pack is recommended by taking the car to a dealer of that brand to run a scan on its capacity and output. That battery health should be considered when negotiating the price you'll pay for the car.

Another consideration is determining what sort of warranty coverage is in play. Most battery packs are covered by 8-year/100,000-mile warranties. Check to see if that coverage still applies and, more important, if it is fully transferrable. Also, the warranty may only be valid if the vehicle has had annual battery health checks required by some manufacturers.

You'll want to know the terms under which the manufacturer will provide replacement. Some warranties may only apply in the case of an outright failure, while others will replace units that have fallen below anywhere from 60 to 80 percent of capacity.

You may also want to check out what sort of drivetrain coverage came with the car. This is usually a warranty separate from both battery and general bumper-to-bumper coverage. These warranties provide for replacement and repair of the electric motor itself, voltage inverters and other electronic controls. Because it's separate from the battery warranty, you need to find out if its provisions still apply and if they're transferrable to a subsequent owner.

Barring a catastrophic battery failure, most of the components like the electric motor and other systems on an electric car are robust and don't require much in the way of routine maintenance. As a result, you have a vehicle that still has plenty of life left in it well above 50,000 miles without much to worry about when it comes to major component issues.

For vehicles with more than 100,000 miles on them, (happy hunting grounds for the true Tightwad), visit the various owner forums online. They will provide you with a wealth of information on potential problems, battery life and how well a drivetrain is holding up on high mileage models.

Used car shoppers are often advised to consider buying a certified pre-owned car from franchised dealers. These CPO programs include a thorough inspection of the vehicle and may come with additional warranty coverage and service packages. Of course, dealers charge more for these cars, citing the peace of mind that comes from the extra vetting and bonus warranty coverage.

Going the CPO route is a smart move for buyers of a conventional gas-powered car, especially if the dealer throws in some free scheduled maintenance. But, since an EV doesn't have the same complexity as gas car and if the original battery and powertrain warranties apply, you're probably better off passing on the extra cost of a certified pre-owned model.

When buying a used EV, either from a dealer or private party, don't forget to ask what else is included with the car. Find out what sort of charging cables, adapters and even plug-in Level 2 Electric Vehicle Service Equipment (EVSE) came with the car as original equipment and ask that all of it be included in the deal. You might be in for a rude surprise once you get home to discover there's no charging cord in the back or that the plug may not fit your particular outlet.

You should also make sure that the original owners' manual and warranty guide are in the glovebox. Both have important information on the care and feeding of your new family member.

Unlike new cars, no two used cars are the same because of their mileage, how and where they were driven, and how well they were maintained. This has a direct impact on the price you'll pay. A new car has a manufacturer suggested retail price (MSRP) and typically you'll pay between that price and dealer invoice.

Used car prices can be all over the map. As a result, there's no hard and fast number on what you should pay for a used electric vehicle, although consulting a price guide, like Kelley Blue Book, will give you a general idea of what to expect to pay.

Bigger picture, here's an overview of the general market for used EVs divided into four categories: Old and Cheap, Moderate Priced in the Middle, Late Model Used and Eclectic Electric Unicorns.

Old and Cheap

These are the first-generation electrics from about 2010 to 2015. There's a wide range of these vehicles available and many are "compliance cars" that manufacturers were subsidizing to generate clean air credits. Some used nickel-metal hydride batteries; others were at the forefront in the push towards lithium-ion chemistry. One thing they all had in common was a range that was generally less than 100 miles. Some of these vehicles also weren't sold nationally, so finding one can be tough outside of California or those places that follow that state's zero emissions regulations.

The first Nissan LEAF has limited range and an odd look.

These vehicles will usually sell for less than $10,000, have higher mileage and will have experienced some battery degradation that may result in a range of 50 or 60 miles instead of the original 80 to 90. Two of these originals are the first-generation Nissan LEAF, sold from 2011 to 2018, and the Mitsubishi iMiEV, a microcar offered from 2012 to 2017. Both have styling that's best described as unusual. The LEAF could go about 75 miles on a charge, while the iMiEV had a range of only 63 miles. A battery upgrade in SV and SL LEAF in 2016 upped range to 107 miles. In any event,

The Ford Focus EV has a range of about 75 miles.

either model is at best a step-up from a road-worthy golf cart.

Boasting slightly more range are the 2012-2016 Ford Focus EV, 2013-2019 Fiat 500e, earlier models of the BMW i3, and 2015-2016 Volkswagen eGolf. The Focus EV's range is about 75 miles.

 The Fiat 500e could go 87 miles on a charge and its small footprint makes it fun to zip around in. While it has a small back seat, for all intents and purposes, it's a two-passenger vehicle.

The BMW i3 offers spirited handling.

The BMW i3 has a funky look with coach-style rear-hinged back doors, and about an 80-mile range. BMW also featured a range-extending two-cylinder gas generator allowing the i3 to travel an additional 100 miles between charges. These models sell for a premium over non-range-extended models.

 The Volkswagen eGolf combined the German-inspired road manners of the conventional Golf with electric power. Kia sold an electric version of the Soul from 2015-2018 in selected states with a range of 93 miles. All these models could be recharged at Level 2 usually in under 6 hours because of their small battery packs. None had the capability of DC quick charging.

Moderate Priced in the Middle

This set includes improved versions of some of the vehicles mentioned above. The BMW i3 would get a larger battery pack boosting range to 153 miles. The same for the eGolf which would, from 2017 on, be able to go 125 miles between charges. The Nissan Leaf had a major redesign in 2018 that gave it more conventional looks and a range closing in on 150 miles. The Ford Focus saw a range bump in 2017 and 2018 models to 110 miles.

The 2016-2021 Hyundai Ioniq Electric falls into this category with its 125-mile range and fast-charge capability restoring 80-percent capacity in just 23 minutes. In 2020 it also received a battery upgrade to 38.3 kWh which boosted the range to 170 miles.

Also joining this group is the 2017 to 2021 Chevrolet Bolt. This model is distinguished by its 238-mile range, which later grew to 258 miles thanks to some engineering changes to the same 62-kWh battery. All Bolts have been subject to a NHTSA recall replacing the battery pack over fire concerns. The longer range of the Bolt has kept resale values up versus lower-range rivals and if you find one that has had the recall work completed, you're

All Chevrolet Bolts were subject to a battery pack recall.

essentially buying a used car with a new battery.

Adding to the value of these vehicles with larger batteries is DC quick charging capability. The net result of improved range and fast charging are significantly higher prices, often above $20,000. Many of these used electric cars have prices that nearly rival those of new entry-level models still eligible for the full $7,500 federal tax credit and other local incentives.

Late Model Used

An emerging market for used electric vehicles will be among those models that the Tightwad might blanch at their sticker prices when new. Given that these used cars don't come with much in the way of incentives, it's not fair to compare them to comparably priced new electric vehicles that do.

But, when you price some of these used EVs with the cost of a similar-sized gas-powered new car, it makes sense to consider them if you're looking at making the switch to electric power. While we covered some new electric cars under $40,000 earlier, these in-between models are still somewhat rare. Perhaps in a few years, there will be an emerging market for used Volkswagen ID.4, Hyundai Ioniq 5 and Kia EV6 models within the reach of the price of a new midsize family car.

For now, the best bang for your buck in getting an affordable electric alternative to a new mid-sized sedan like a Toyota Camry or Honda Accord in the mid-to-upper $30,000 range would be a used 2013-14 Tesla Model S or 2018 Tesla Model 3.

Eclectic Electric Unicorns

While we're focusing on finding mainstream affordable electric cars, there is a Tightwad subset that might want something a bit more exotic on a budget. Of course, the first-generation LEAF, Mitsubishi iMiEV and Fiat 500e are among those that fit the bill.

But there are even rarer birds out there. These models were sold in limited numbers in select states. While these vehicles are affordable transportation alternatives, there's the added appeal in that you won't see them on every corner. Some, because of their lease-only programs by manufacturers or limited state availability, are even more rare.

Among the limited-state models to look for is the 2014-2016 Chevrolet Spark EV. Before the Chevrolet Bolt was introduced, GM sold an electric version of the subcompact Spark in California, Oregon and Maryland. Limited in range to 82 miles from its 21.2 kWh battery, the Spark never sold in significant volumes. The car's small footprint was a good match for an electric powertrain that

The Spark EV preceded the Bolt but was sold in only three states.

delivered 140 horsepower to the front wheels. The electric Spark is known for delivering superior performance over its gas-powered counterpart.

Mercedes-Benz also dipped its toe into the electric waters with two interesting models: the smart fortwo ED (for electric drive) and the B-Class, sort of a cross between an SUV and a station wagon. The smart fortwo ED is a tiny city car with room, as the

The two-seat smart fortwo electric is at best a step-up from a golf cart.

name implies, for just two. Originally introduced in 2011 under a demonstration lease program with 250 units, the car eventually became available for sale in both a coupe and convertible body style. Its limited range and utility resulted in few sales before the model was discontinued in 2018. Prices are likely to be less than $10,000, range will be extremely limited.

The five-passenger Mercedes-Benz B-Class Electric Drive was introduced in 2015 and was only sold for three model years. Its 36-kWh battery pack was good for 85 miles range. It was sold primarily in states following California emissions, but it also was available nationally on special order.

Another so-called "compliance car," the 2013-2014-Honda Fit EV was leased in two states, California and Oregon. Like other subcompact electrics of this era, the battery was small and the range about 80 miles. Like the Chevy Spark, the electric drive exponentially boosted the Fit's fun-to-drive factor. Honda also sold an electric version of its mid-size Clarity from 2017 to 2019.

Equipped with a 25.5-kWh battery, the $36,000 Clarity electric could go 89 miles between charges. As part of a three-model lineup that also included a plug-in hybrid and fuel cell electric versions, this EV sold only in California and Oregon.

Will There Be "Beater" EVs?

The late model used electric vehicle market, especially with the availability of manufacturer-sponsored Certified Pre-Owned programs, will mirror what's going in the current second-hand market for gas-powered vehicles. Many of these electric cars, especially those with lower mileage and ranges that exceed 200 miles, will see strong prices especially if gas remains high.

What we don't know is if there will develop a market for high mileage, cheap "beater" electric cars. Many affordable gas-powered beaters are sold by independent dealers. While there is a lot of talk about recycling the battery packs of EVs, not much has been written about what to do with the car part, especially if the body and electric motor are in relatively good shape.

Perhaps there's a market for either refurbished or even new battery packs that could give these vehicles a second life at a much lower cost than either a new or late model used EV. Much of it depends on the cost of either refurbishing existing battery packs or producing new ones at a relatively affordable price.

At current battery pack prices, which can run $8,000 to $12,000, it's not likely to happen. A remanufactured battery pack at $2,000 might work. It's an opportunity that could open new doors for a national used vehicle chain like CarMax or some new scrappy competitor looking to serve a market for low-cost electric transportation. In any event, you'll know that the electric vehicle has truly arrived if there are $5,000 beater EVs on the market.

Tightwad Pro Tip:

While an electric vehicle makes a great short commuter or local errand runner, a great use is a first car for a high school student, especially if you buy one with limited range. You know they'll never be able to go farther than half that distance from home, the top speed at best for most electrics is about 90 to 100 mph and you don't have to give them a gas card which they will use to buy snacks and other treats when otherwise filling the tank.

8. GETTING A CHARGE INTO IT: HOME EDITION

The big appeal of driving an electric vehicle is that it's like having a filling station in your house. While you may not have the same selection of snacks and cold beverages on hand, at least you can count on clean bathrooms. The convenience of being able to recharge your electric vehicle does come at a price, not just in dollars, but in time. And any steps you take to reduce that time will cost you more money. But perhaps not as much as you think.

Level 2 home charging the Nissan LEAF.

Electric vehicles are capable of being recharged off a home's standard 120-volt outlet. However, as we have seen, it can take an eternity to replenish a battery pack that offers as modest a range as the LEAF's 149-mile distance. After driving around and exhausting the initial charge on my LEAF S, I plugged it into a 120-volt garage outlet with the handy adapter that came with the charger cord. With just six miles of range left, the LEAF took about 30 hours to fully recharge.

In comparison, using a Level 2 system at 240 volts can do the trick in just over seven hours. Early adopters of electric vehicles looking to access Level 2 had charging systems installed in their garages, sometime at a cost as high as $2,000. Fortunately, the price has come down with units on the market costing as little as $500-$1,000 installed. The high-end setups usually have Wi-Fi capability that will allow you to schedule sessions, set charging levels and receive notifications of the state of charge via your smartphone.

The Tightwad can take heart – those more elaborate and expensive Level 2 home charging stations aren't really needed. Manufacturers incorporate the Level 2 electronic controls needed to communicate with the vehicle into the charging cable itself. On the window sticker, look to see if the portable charger cable that come with the car has a 120V/240V EVSE (Electric Vehicle Service Equipment) designation. All you really need is a 120- or 240-volt outlet nearby to provide current.

In places like California where the garage essentially serves as a basement/laundry room, you'll often find a 240-volt outlet rated at 30 amps for use with an electric dryer. EVs are generally equipped with 240-volt plugs that will allow them to connect to more powerful 4-prong 50-amp outlets. You can hire an electrician to come out and put in a higher amp circuit breaker and outlets that match the plug, or, if you don't mind a slightly lower and slower rate, you can buy a plug adapter, as I did, for about $35. The adapter works fine since, in the case of the LEAF, it recharges at a maximum 27.4 amps which more closely matches my available outlet's rating.

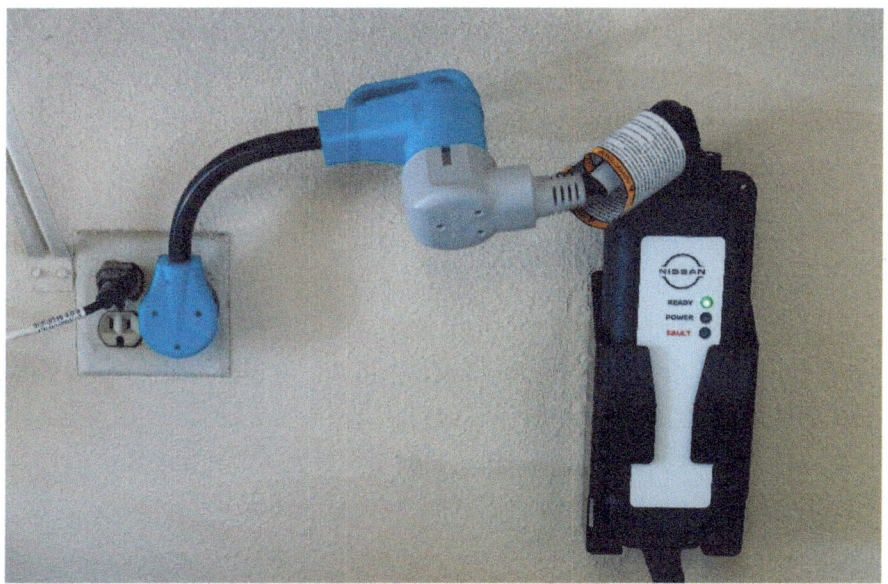

The LEAF comes with a charging cord for both 120- and 240-volt service. Here it's plugged in to a standard 240-volt outlet using an adapter.

In addition to being able to Level 2 charge our LEAF, I have the benefit of using the same cord on our Hyundai Ioniq plug-in hybrid. When using the standard 120-volt cord that came with that car, it takes about six hours to get the full 29-mile range. At Level 2, the process takes half that time.

Having a garage, especially one already with a 240-volt outlet, is a huge advantage. Getting an electrician to wire your garage for one is an added charge. With the proliferation of aftermarket plug-in Level 2 chargers, you can easily find one that fits your budget with the features that make the most sense for your needs, whether it's a simple recharge or being able to program sessions to include pre-conditioning cabin heating and cooling.

But if you don't have access to a garage, say you live in an apartment or condo, keeping down the cost of electric car ownership becomes more of a challenge. If you have access to charging at work where employers either subsidize the cost or give it away for free, you probably can make the EV experience work for you. If you rely solely on public charging, it can cost considerably more than at-home charging as we'll show you in the following chapter.

The other option, especially if you live in a condo or development with a homeowner's association, is to ask the HOA to install community charging stations. These kiosks, in which the HOA will get a revenue share and may also be able to set or subsidize rates, generally offer Level 2 charging. If your association has the money, ask them to add Level 3 DC fast charging.

As for putting in Level 3 fast charging at home, be prepared to spend a lot of money. Like $50,000. That's not exactly in the Tightwad's wheelhouse.

How Often Should I Charge?

The other question that comes up with home charging is whether you should charge your electric car each night or only when the range drops down to 20 percent. It's analogous to topping off your gas tank daily or filling it once a week when that little orange light comes on (and you think you can go about 30 miles more).

That low-fuel or state-of-charge warning psychology is an interesting contrast between conventional and electric car ownership. With the ability to refuel quickly, some drivers are willing to run the risk of running out of gas. But with an electric, range anxiety is a real issue. Some may find themselves freaking out at any remaining range less than 30 miles. It's understandable because you really can't buy a can full of electricity to pour into your battery at the side of the road. Run out of charge, and it's time for a flatbed tow. So, there's a comfort level built into beginning your day in your electric car with a full charge on board.

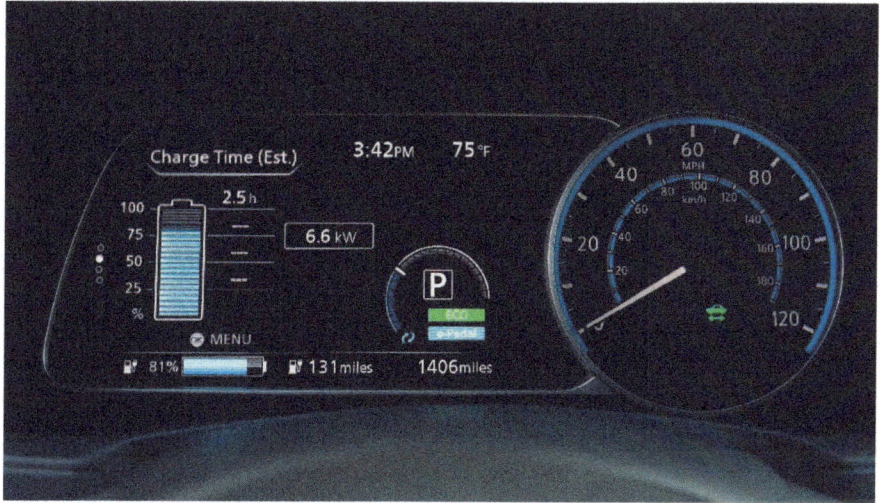

This readout gives state of charge and an estimate of recharging time.

When it comes to charging every night, a lot depends on how far you drive every day. If you have an average commute of 15 miles each way and do a few errands, you probably could go every other day with an EV packing a range of about 200 miles. However, if you're particularly anxious and need to top off daily, you're best setting an upper limit on the battery state of charge at perhaps 80 or 90 rather than 100 percent.

Recharging a battery is like blowing up a balloon. It takes a lot of effort to get those first few puffs in, then it's relatively easy to blow up, and then gets harder when it's near full. As you reach capacity, the air inside a balloon pushes back. The same theory applies to a battery. It requires a lot of energy to start charging a fully depleted battery, gets easier between 20 and 80 percent capacity, and then gets harder again for that last 20 percent of charge. The resistance encountered outside that sweet spot can raise temperatures and degrade the battery cells.

When General Motors was having issues with its Bolt battery packs catching fire, it advised owners not to charge the battery to more than 90 percent capacity.

That's also why there's a school of thought over repeated DC fast charging having an adverse effect on battery life. The electricity flows in at a higher rate and generates more heat which, when done repeatedly, can cause a battery to degrade and slowly lose capacity.

It's worth noting that whenever fast charging stats are quoted, the time to charge represents only an 80 percent fill. The time of the session can nearly double to fully recharge the battery to its 100 percent capacity.

Cycles Count

Whereas mileage is a wear factor on an electric vehicle's moving parts just as it is on a gas-powered car, a more critical element is the number of charge/discharge cycles that the battery pack undergoes. You'll get more cycles out of a battery if it's recharged between 20 and 80 percent routinely at Level 2 than one that has a lot of cycles that push it to a full charge. The bottom line here is finding your comfort level with how far you travel each day, what sort of range is left when you come home and how often you really want to take the effort to stick the plug in your car. I know it's simple, but hey, even the Tightwad can be lazy.

To recap, here are some charging guidelines recommended by BatteryUniversity.com:

Limit ultra-fast charging, especially when the battery is cold. Use Level 2 when possible.

Only charge the battery to the level needed for the daily routine. Full charge hastens capacity fade.

Do not discharge the battery too low as this increases the internal resistance. Charge more often.

Charge and use the battery at room temperature. Operating when cold reduces capacity.

Store the battery in a cool place at partial charge. Usage and storage have different requirements.

Moderate the battery to room temperature in winter before charging and driving. The Battery Management System (BMS) may do this automatically.

Charge the EV after a sabbatical. Resting at low charge reverses capacity fade.

It is best to let the battery rest at low State of Charge (SoC) and only charge before use. Dwelling at low charge reduces calendar aging and may also reverse capacity fade.

What It Costs to Charge at Home

The cost depends on how far and fast you drive, and how much you pay per kWh. The last depends on a lot of factors including if you have tiered or time-of-day rates, and where you live. The national average cost for electricity is 13.75 cents per kWh, according to the U.S. Energy Information Agency (EIA). It ranges in the Lower 48 states from as high as 24.03 cents per kWh in Massachusetts to as little as 9.35 cents in North Dakota. California comes in a close second for the highest rates at 23.72 cents, according to the EIA. The most expensive electricity is in Hawaii at over 35 cents per kWh.

A quick way to calculate your cost is to look at the EPA rating on the window sticker. The 2022 Nissan LEAF S is rated at 30 kWh of electricity per 100 miles, or 0.3 kWh per mile. Using California's average price of nearly 24 cents per kWh, it works out to about 7.5 cents per mile. If I had a similar compact hatchback that gets 30 mpg, at $5 per gallon the per-mile cost for fuel is more than twice that at 16.6 cents per mile. At $2.50 a gallon, it's a push. So, while there is definite savings in electric versus gas power, the savings aren't as dramatic as you'd think. Also, you should take into consideration whether you're charging your vehicle off-peak or peak — in California, peak rates can be as much as 46 cents per kWh, which closes that gap considerably.

But just as your mileage may vary, so do your electric rates and local gas prices. Cheap hydroelectric power in the Northwest holds down electric rates, which, given the high cost of fuel on the West Coast, works out in favor of going electric. Oil-producing states like Texas and Oklahoma have fairly low electricity rates and yet, also enjoy pump prices that are among the lowest in the nation. The spread-out nature of those two states and low cost of fuel may favor gas-powered vehicles. So, wherever you live, find out your residential power rate versus the price of a gallon of gas and do the math.

Also keep in mind that oil, a global commodity, can see dramatic price swings. In 2008, the average price of gas was $3.27 per gallon. By 2015 it was $2.15. Now it's up over $4 nationally. In addition to being able to boost domestic production, the U.S. can also import lower cost oil from overseas. But other than getting some power from Canada, we really can't import electricity. To get more juice to power millions of electric cars, the nation needs to invest in additional generating capacity and a more robust grid. All this additional investment will likely be reflected in higher power rates.

Generating costs, which the EIA says account for up to 56 percent of the total price of electricity (the rest is in transmission and distribution) rely on diverse sources that include coal, oil, natural gas, and nuclear power. Renewables include solar, wind, hydroelectric and geothermal. The EIA also reports that electric prices are up 4.3 percent since 2020, the largest increase since 2008, largely due to the rising price of natural gas. In an inflationary environment, all energy is going to cost more. Smart Tightwads will take into consideration all these factors, crunch the numbers and do whatever takes the smallest bite out of their bank accounts.

While electric vehicle ownership avoids the high cost of gas prices, you might not be getting scot-free from paying some sort of road tax. The federal gas tax is 18.4 cents per gallon. State rates range from as low as nine cents per gallon in Alaska to 51.1 cents in California. The U.S. Department of Energy reports that the average motorist paid between $141 and $398 in annual gas taxes as recently as 2019.

Some states are proposing some sort of vehicle use fee based on annual mileage that would capture some of this revenue that electric cars avoid by not having to fill up. In the meantime, 30 states require special registration fees for electric vehicles and 14 also assess additional fees on plug-in hybrids, according to the

National Conference of State Legislatures.

These extra annual fees, which are on top of the standard new vehicle registration, range from $50 per year in Colorado, South Carolina, and Hawaii to as much as $225 for pure electrics in Washington state. Alabama, Arkansas, Ohio and Wyoming charge $200. Fees for plug-in hybrids range from about $50 in Iowa to $100 in Alabama, Arkansas, Ohio and West Virginia. Some states are also indexing these fees to inflation, so expect these costs to continue to go up. It goes to show that two things in life are inventible: death and taxes, for EV owners and Tightwads alike.

The Home Solar Question

So, what about home solar? If you already have a system, getting an electric car can be a no-brainer. Of course, that's if you have net metering versus net billing. With net metering, you get a credit of one kWh for every one you produce during the day that's returned to the grid. Net billing, on the other hand, credits any solar electricity your home system generates at the wholesale rate, while continuing to bill you for any electricity you use from the grid at a retail rate.

This is important because if you're like most people who are out and about during the day, you usually recharge at night during off-peak hours at the lowest rate. In this case, you'll be using power from the grid and not your solar system (unless you pay for expensive battery storage to capture some of that sunlight-generated energy during the day). In any event, you'll still be paying for your electricity, but at a much lower rate than those who don't have solar.

If your goal is to buy an affordable electric car and want to install solar, then stop right there. Adding solar panels to your house can cost $20,000 or more. And while there are federal investment credits for home solar, they are being phased out and will expire in January 2024. State and local credits still may still apply.

Even with the fuel savings of lower priced electricity over high priced gasoline, it would take years to amortize the initial price of going solar just to charge your electric car.

The other consideration is sizing your new solar array to not only fit your home and transportation power demand, but also appropriate to the available sunlight in your area. For those looking to go off the grid entirely, you'd need an expensive home battery system to be able to utilize the harvested solar energy around the clock. A Tesla Powerwall battery pack with 13 kWh of storage costs about $7,500. The average home consumes about 30 kWh of power per day. If you drive 30 miles a day in a car that's rated at 30 kWh per 100 miles, you'd be adding at least another 9 kWh to the total. As a result, you'd need three of these home storage batteries or another $22,500 worth to go off grid.

Trying to recharge your electric in real time just using solar panels alone would require a much larger and more expensive setup. Solar power is like collecting rain for your household water needs. You wouldn't get much water pressure while the rain is coming down. Instead, you need to collect it over time, store it and then use its mass and gravity to generate the pressure you need to make it suitable for everyday use.

The same applies to an electric vehicle. If it were plugged into your solar system and was drawing at the rate it normally would from the grid, you'd soon overwhelm a normal 8-10 panel system's ability to keep up. Solarnerd.com estimates that the Nissan LEAF with its 6.6 kWh on-board charging capability would need 44 400-watt solar panels to generate the 17,600 watts needed for charging. The website concludes that "trying to achieve off-grid electric car [charging] is really expensive and impractical and doesn't make any sense from either an environmental or economic perspective."

Just as we continue to wait for the next battery "breakthrough" that will lower costs, reduce the time it takes to recharge, and extend the range, so too we wait for the next generation of solar panels that will boost efficiency and output while lowering costs.

Tightwad Pro Tip:

Check to see if your local utility has special rate plans for those who own electric cars or plug-in hybrids. If you qualify, you may see substantial savings in rates, especially when recharging your vehicle during off-peak periods.

9. GETTING A CHARGE INTO IT: TRAVEL EDITION

As electric vehicle batteries improve, range anxiety should ebb by being able to travel longer distances between recharging. Those early models of a decade ago were barely bumping up against the 100-mile marker, and now you can find entry-level EVs with the ability to go anywhere from 150 to 260 miles on a single charge.

As range anxiety fades, it may soon be replaced by charging anxiety. This is the worry about finding a public charging station and whether it's occupied or even working in the first place.

J.D. Power's U.S. Electric Vehicle Public Charging Study reports that access to public charging "is a key component in the overall adoption of electric vehicles by the broad population." Brent Gruber, senior director of global automotive at J.D. Power, adds, "Unfortunately, the availability of public charging is the least satisfying aspect of owning an EV. Owners are reasonably happy in situations where public charging is free, doesn't require a wait and the location offers other things to do – but that represents a best-case scenario. The industry needs to make significant investment in public charging to assure a level of convenience and satisfaction that will lure potentially skeptical consumers to EVs."

The Power study also found that among electric car owners, the most often cited problem related to public charging was finding an outlet that works. A total of 58 percent of the survey's respondents cited an instance of finding a charger that was out of service. That compares with 14 percent who found all the chargers occupied or had a long wait to recharge.

The lack of public charging infrastructure, particularly DC fast charging, is cited as an obstacle to widespread EV use for long-distance travel. Gas- or diesel-powered vehicles still retain an advantage when it comes to the convenience and ubiquity of gas stations and the short time it takes to refill a fuel tank.

That speed of refilling also means that a gas station can have greater throughput from even a limited number of pumps. The 115,000 gas stations in America today service the more than 280 million cars on the road. The way these stations can accommodate such a huge number of vehicles is a combination of the relatively short time it takes to refuel and the fact that most stations have upwards of 12 pumps per station, which means there are more than a million pumps available at any given time.

It's a different story when it comes to electric vehicles. According to the Alternative Fuels Data Center maintained by the U.S. Department of Energy, there are 45,973 public charging stations in the U.S. with just over 114,000 charging outlets (each the equivalent of a gas pump) across the country. Of that number, about 92,000 are Level 2 chargers while the rest are DC fast charging outlets. Currently, there are just under two million full electric vehicles on U.S. roads.

A mitigating factor, of course, is that ability to charge at home. Few if any motorists have a gas pump in their garage. Still, as overall EV penetration grows rapidly to include those who don't have the ability to charge at home, building a larger public charging infrastructure is imperative. EV charging stations may not need to have as many fast chargers as the average Buc-ee's has pumps, but these facilities should have more than the two or three chargers that are the norm today.

ChargePoint kiosks offer both Level 2 and DC fast charging.

The federal government's recent Infrastructure Investment and Jobs Act sought $15 billion to establish 500,000 stations but ended up earmarking $7.5 billion to build half that number. Most will be in urban areas, airports, apartments and offices, while 2,000 will be built along highways and in rural areas.

These new facilities will be in addition to other public networks maintained by Tesla (which has announced plans to open its network of charging stations to all EV owners) as well as private concerns like Electrify America, ChargePoint and EVgo.

Electric car charging stations won't necessarily have to be as ubiquitous as our gas and diesel networks, but they will have to be able to accommodate more vehicles at a single time to reduce the wait for charger access. While individual charging networks like Tesla have apps that allow for reserving a spot, there may need to be a universal software scheduling program that would allow electric vehicle owners to reserve a slot at any number of charging locations. This would ease some of the planning and logistics needed in electric cross-country travel.

Public Charging Prices a Concern

When it comes to charging, the best rate you're going to get will be at home, although there are some public charging outlets that are free – offered as a perk at workplaces, retail outlets and municipal facilities. Most other locations are owned and/or leased by the property owner, and each service provider is allowed to set rates and share in the revenue. While some businesses or workplaces may continue to offer free Level 2 charging, as electric vehicles become more numerous, these charging stations, especially those offering fast DC service, will become profit centers in their own right.

As for charging away from home, the Tightwad believes that free is good if you can find it. But often you'll be paying half again as much as you do at home for Level 2 service. In some states, unlike your home service that's billed a set kWh rate, you may be charged at a per minute rate for the session rather than per kWh, which could cost more.

And that goes double for DC fast charging. The investment required to provide a high voltage, quick charge outlet is high, so there's no freebies here, either. While some fast charging may be based on a similar kWh charge as Level 2, you'll also encounter rates based on the time it takes to charge or session rates that can add a $10 fee in addition to the cost of the electricity.

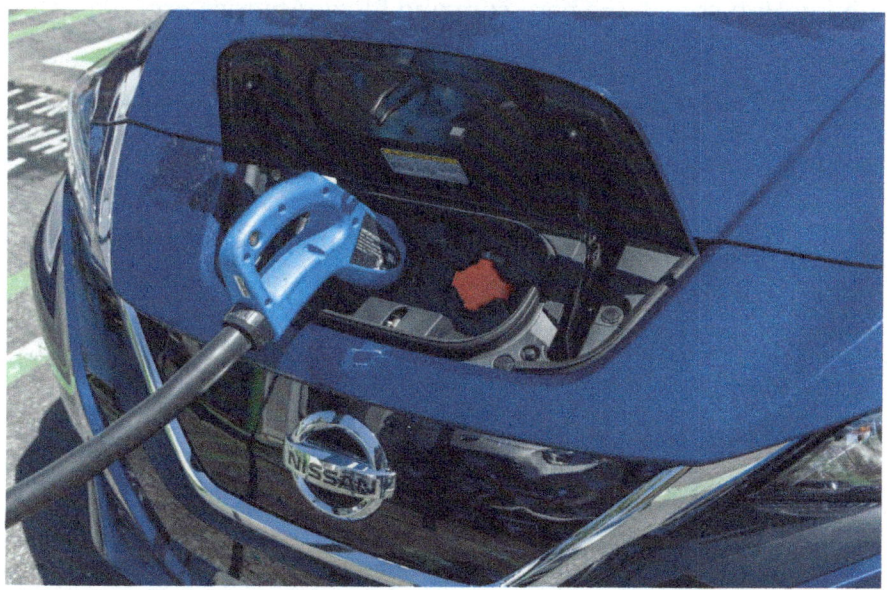

DC fast charging the LEAF. The larger outlet supports a specific plug for the task. Some newer plugs incorporate both Level 2 and Level 3 connections in the same unit.

Other fees to be wary of are idle charges. Once your vehicle is recharged, you might be hit up for the time your vehicle remains in the space. These fees can range from 25 cents to over $1 per minute. As a result, you're essentially paying anywhere from $15 to more than $60 per hour for parking.

If you can't find free charging, there are two public charging options: pay-as-you-go or subscription services. Some charging providers combine the two with free or nominal membership fees plus a minimum credit card balance requirement. You pay as you use their service, often at a discounted rate from what non-members pay. Monthly subscriptions have even deeper discounts.

How to Find Public Charging

Map-based charging station directories are available for free on ChargeHub.com, PlugShare.com and PluginAmerica.org. Here's the list of the major and regional public charging operations from the U.S. Department of Energy's Alternative Fuels Data Center:

Blink

With 1,367 locations and 3,308 ports located throughout most of the United States, Blink works with private businesses to host charging stations that feature both Level 2 and Level 3 DC Fast Charging. Also, Blink produces a line of home Level 2 chargers. Membership is free and provides discounts of 20 percent or more when charging at the company's locations. Rates vary by location, ranging from 39 to 79 cents per kWh. In states that don't allow kWh charging, the rate is two to three cents per 30 seconds. In California, members are charged 49 cents per kWh for Level 2 and 59 cents per kWh for DC fast charging, while non-members pay 10 cents more per kWh. In some states like Texas and Pennsylvania, fast charging costs $6.99 per session for members and $9.99 for non-members.

ChargePoint

The largest public charging network in the country is maintained by ChargePoint with 26,402 locations and 47,482 ports across the country. The free ChargePoint app includes all available charging stations (including free outlets), rates and availability. ChargePoint locations have rates that vary since it allows the business that partners on the installation to set them. In addition to selling the electricity, some locations will charge a parking fee if you leave the vehicle parked in the space beyond two hours. ChargePoint also retails Level 2 home charging units to the public. There's no charge to sign up for a ChargePoint card, however, when you enter your billing information and use the card for the first time, the company adds $10 from your credit card and will replenish that charge whenever the balance dips below $5.

Electrify America

This growing network is open in 772 locations with 3,433 ports nationally. It's free to join and download the app. Guests and pass members are charged the same rates, while Pass+, which costs $4 per month through the app, provides a 25-percent discount on charging. The kWh charges, where applicable, are the same whether you use Level 2 or Level 3 Fast Charging. In California, the company charges 43 cents per kWh, while Pass+ members pay 31 cents. In states like Pennsylvania and Texas, the rate is 32 cents per minute for fast charging and 16 cents per minutes for Level 2 charging, with Pass+ members paying a respective 24 and 12 cents per minute. You might not want your car to linger long after charging; the fee is 40 cents per minute, which works out to $24 per hour for parking.

EV Connect

With 743 locations and 2,994 ports, EV Connect is largely a business-to-business supplier of charging stations for use in retail outlets, offices, apartment buildings and other public venues. Among its clients are hotel chains Days Inn and Marriott, the Panda restaurant chain and several utilities including New York Power Authority and Southern California Edison. Like other networks, EV Connect lets its partners set rates and terms. EV Connect has a consumer facing app that helps locate charging locations, determine charger availability and lists rates.

EV Gateway

EV Gateway is another provider of electric vehicle charging at 27 locations with 132 ports. Like others in this category, it allows the business client to set rates and terms. EV Gateway offers an app to help locate charging stations and facilitate payment. There is an auto reload feature that will automatically add a $10 credit to the account from your charge card when the balance drops below that level. Their software includes color notification of a charging site's availability: green is open, blue is in use and gray means it's unavailable.

EVgo

This fast-growing network includes 869 locations and 2,214 ports (including over 800 fast charge outlets) in 34 states. In addition to working with Nissan, EVgo has inked deals with General Motors, Toyota and Subaru. In California, rates range from 29 to 39 cents per kWh (prices can be higher depending on time of use). Guests must pay a $1.99 per session fee while both guests and members pay $3 to reserve a charging slot. Members also must maintain a minimum $4.99 balance in the account.

EVgo Plus memberships are available at $6.99 per month and feature rates that are generally 10 cents per kWh cheaper. EVgo Plus members pay nothing for reservations. All members are eligible for a rewards program that pays five points for each dollar spent charging. Redeeming 2,000 points earns $10 in free charging.

Flo

A subsidiary of Quebec-based AddEnergie, Flo is a charging network that has spread from Canada into the U.S. with 267 locations with 491 ports primarily in the Northwest and Northeast. It recently integrated its operations with ChargePoint to ease access to its network. To boost its visibility, Flo has become the official electric vehicle charging partner of Major League Soccer's New York Football Club.

FPL EVolution

Florida Power & Light plans to build up to 600 locations in the Sunshine State, expanding its current base of 17 locations and 86 ports. To access the system, users need to download the FPL EVolution app. Rates are set at 30 cents per kWh with a 40 cents per minute idling rate for leaving the car at the charger for more than 10 minutes after a session has ended.

Francis Energy

Also regional in nature, Tulsa-based Francis Energy aspires to grow beyond its 117 locations and 630 ports in Oklahoma. It recently announced its expansion in Kansas and will be offering facilities along the I-35 and I-70 corridors. Its business model includes a plan to build recharging locations every 50 miles along major highways in middle America. Francis offers both Level 2 and DC fast charging through its app.

Greenlots

Acquired by oil giant Shell in 2019, Greenlots operates 1,092 locations nationally offering 2,937 charge ports. While the charge stations may be still identified by the Greenlots brand name, the consumer app is called Shell Recharge. The app is also connected to other providers including ChargePoint, Flo and EV Connect. While there is no membership fee for the app, recharging prices vary by location.

Livingston Energy Group

Located in Upstate New York, Livingston Energy Group is listed as having 29 charging facilities with 188 ports. The company works with businesses, residential units and offices to provide charging infrastructure. It has no consumer facing app for use or billing.

OPConnect

Based in Portland, Oregon, OPConnect has 149 locations and 564 ports in various locations in Oregon, Washington, California, and Hawaii. Its privately maintained network is open to the public and members who have downloaded the company's app. Some host sites may offer free charging, while other locations have an hourly rate that varies from $1.25 to $3 for Level 2 and $5 to $7 for fast charging sessions.

PowerFlex

With 60 locations and 907 outlets, PowerFlex is a California-based provider of electric charging kiosks. It has its own consumer app that allows users to find and pay for charging sessions at PowerFlex sites. Rates and idling charges vary by location.

SemaConnect

Founded in Bowie, Maryland, SemaConnect is a Level 2 charging network with 1,982 locations and 5,995 ports maintained by businesses, residential communities and workplaces. In addition to providing the charging stations, it maintains a free app for finding stations and billing purposes. A $10 credit is required when first using the app; subsequent account replenishment starts at $20, but users can load up to $200 in charging credits. Rates are set by the charge site operator.

Tesla

The second largest charging network in America today is maintained by Tesla. The company operates 4,400 locations and 11,243 ports as part of its Destination network, plus 1,292 Supercharger stations with 13,072 ports that feature DC fast charging. While Elon Musk tweeted that the network may be opened to non-Tesla electric vehicles, details on how non-Tesla owners can use the facilities are still forthcoming.

Volta

Using stylish kiosks that deliver messages from clients, Volta offers a unique ad-supported approach to offering free charging at 996 locations with 2,374 ports. Most of Volta's stations are Level 2, but it has begun installing Level 3 Fast Charging that also is free for the first 30 minutes or 150 miles of charge. To find stations and charging availability, Volta has a free consumer app.

Tightwad Pro Tip:

When shopping for an affordable electric car, research your public charging options. Volkswagen provided the seed money to establish Electrify America. Buyers of VW ID.4 get three years of free DC fast charging on that network. Other manufacturers may have similar deals with other charging network providers. Nissan gives $250 in charging credits at EVgo.

10. EV MAINTENANCE, SAFETY AND INSURANCE

One of the benefits of owning an electric car, from the Tightwad's perspective, is no longer having to collect or clip coupons for oil change discounts. Less routine maintenance not only saves money but also the time spent either taking in your car for work or doing it yourself in the driveway.

It also gives you another reason to ignore those annoying robocalls about your car's warranty. Still, an electric car is a car and there are parts of it that are the same no matter what's under the hood. You must pay attention to tire and brake wear, although, thanks to regenerative braking from the electric motor, the latter is less than on a conventional car.

In addition to making sure your tires are inflated properly (low pressure can increase rolling resistance, increase wear and reduce range), other routine tasks include keeping the windshield wiper, coolant and brake fluids topped up, replacing any worn wiper blades and burned-out light bulbs, and lubricating door, hood and hatch hinges and latches. The coolant is used as part of the vehicle's climate controls. On the LEAF, for instance, there is a reservoir access cap located under the hood, even though the original factory fill is expected to last 125,000 miles and 15 years.

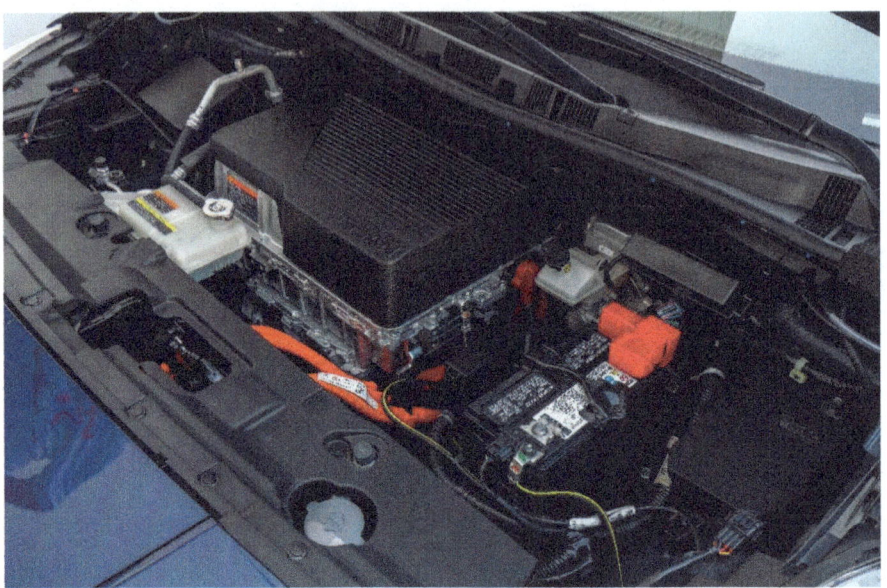

A 12-volt lead acid battery (foreground) runs accessories. At some point, it may need to be replaced.

Most electric vehicles also have a separate 12-volt system that operates accessories like the infotainment system, heating/air conditioning fan, lights, heated seats, power door locks, seat adjustments and windows. That battery is automatically recharged when the vehicle is either plugged into the wall or fed current from the motor and battery pack when the car is in motion. However, over time it can degrade and eventually may need replacement.

Still, the number of items that need regular attention are fewer on an electric vehicle. We Predict, an analytics company in Ann Arbor, Michigan, released data pointing to the fact that "EVs have fewer mechanical parts than gasoline vehicles, contributing to 22 percent lower repair costs. The primary difference in cost, however, is maintenance. The average amount spent on maintenance per EV in the first three years is $77, significantly lower than the $228 average for gas vehicles."

Manufacturer Recommendations

So, what does the manufacturer's recommended service look like on a vehicle that doesn't require the same level of attention as a traditional gas-powered vehicle? Many manufacturers, like Nissan with the LEAF, still advise owners to stick to the same 7,500-mile/6-month maintenance intervals that they have for their gasoline-powered models. These visits typically include inspections of most systems and tire rotation. Every 15,000 miles or once a year, Nissan recommends replacement of the in-cabin air filter. Annual battery pack checks are included in the first two years and may cost from $40 to $120 thereafter. Nissan also requires an annual battery health test for its 8-year/100,000-mile battery warranty to remain in effect.

Brake pad and rotor life may be longer on an electric vehicle, but brake fluid service is another matter. Nissan recommends a brake fluid replacement at 60,000 miles or four years on its standard cars. However, on the LEAF, it calls for it to be changed every 30,000 miles or even as soon as 15,000 miles in what Nissan characterizes as severe use. This includes repeated trips of less than five miles, rides of 10 miles or less in freezing temperature, hot weather stop-and-go driving, low speed driving and idling for long distances, operating in dusty conditions, use on rough, muddy or salt-treated roads, or regular use of a car-top carrier.

Replacing brake fluid can run about $300. One of the reasons for more frequent service is the car primarily relies on regenerative braking, so the hydraulic brakes aren't as heavily used. Under less pressure because of this light duty cycle, the fluid in the lines may attract excess moisture that might cause brake line and component corrosion.

Another fluid that requires monitoring is the reduction gear oil in the electric motor. Regular maintenance includes inspection for signs of leakage.

At 45,000 miles, Nissan also recommends that the LEAF's intelligent key fob's battery be replaced.

As for repair costs, online cost estimator RepairPal says "the average annual repair cost is $748 which means it has average ownership costs. The severity and frequency of repairs are much lower than other vehicles, so the LEAF is one of the more reliable vehicles on the road."

A broader Consumer Reports study in 2020 put lifetime maintenance and repair costs for the average electric car at $4,600 compared to $9,200 for an internal combustion vehicle – respective costs of three and six cents per mile.

Collision Repair Costs

As for crash repair costs, most of the affordable electric cars on the market today use conventional approaches to their construction. In the case of purpose-built electrics, like the LEAF and Chevrolet Bolt, they are similar in appearance and functionality to compact hatchbacks. Others, like the Hyundai Kona and Kia Niro, share their architecture with hybrid and gas-powered variants. Their unit-body construction is made of steel, which means that in most crashes the body panels can be either replaced or repaired at about the same cost as a standard vehicle.

The only tricky part of accident repairs is if the battery pack becomes compromised in any way, which can result in the need for replacing what is the most expensive part of the vehicle. Higher-end electric vehicles may employ more exotic body panels made of composites or aluminum, which are far more expense to fix than vehicles using steel.

Which brings us to another topic of concern: safety. Like all cars on the road today, electric vehicles must meet federal safety standards. All these vehicles are crash tested by their manufacturers to determine if they pass muster. You'll find the same standard safety equipment, including front and side airbags, as well as built-in measures like side impact beams, front and rear crush zone protection, and rollover protection from the pillars supporting the roof.

Other safety features include rear back-up cameras, tire pressure warning systems, and automatic driver and safety assists. The assist features, including front collision warning with automatic emergency braking as well as lane departure and blind spot alerts with lane keeping and avoidance assist, while not required by law, are becoming standard equipment on all vehicles.

Battery Pack Safety

Manufacturers build in additional protection for the battery packs. The Alternative Fuels Data Center says "battery packs are encased in sealed shells and meet testing standards that subject batteries to conditions such as overcharge, vibration, extreme temperatures, short circuit, humidity, fire, collision and water immersion. Manufacturers design these vehicles with insulated high-voltage lines and safety features that deactivate the electrical system when they detect a collision or short circuit. EVs tend to have a lower center of gravity than conventional vehicles, making them more stable and less likely to roll over."

The vehicles are also equipped with cut-off switches that, in case of a crash or rollover, disable the electric system and in the process isolate the battery pack.

First responders are being trained how to identify high voltage lines and use the latest techniques in extinguishing the flammable electrolyte used in batteries. While these materials are not as volatile as gasoline, electrolyte fires can appear to be extinguished only to flare up later.

Just as gas powered cars can catch fire in an accident or during refueling, the same applies to electric vehicles. The National Highway Traffic Safety Administration (NHTSA) states it "does not believe that electric vehicles present a greater risk of post-crash fire than gasoline-powered vehicles. In fact, all vehicles – both electric and gasoline-powered – have some risk of fire in the event of a serious crash."

The safety agency further says that "However, electric vehicles have specific attributes that should be made clear to consumers, the emergency response community, tow truck operators and storage facilities. Out of an abundance of caution to prevent injury and loss of property, the interim guidance identifies considerations and actions for all electric and hybrid-electric vehicle crashes, including those involving the growing number of vehicles powered by lithium-ion batteries."

If the electrolyte from a lithium-ion battery ignites, NHTSA tells fire departments that it will "require large, sustained volumes of water for extinguishment. If there is no immediate threat to life or property, consider defensive tactics and allow fire to burn out." The safety agency also points out that there is a risk of reignition of a battery pack fire that may seem to be extinguished.

The NHTSA has issued specific advice regarding damage or fire involving an electric vehicle. This guidance includes:

Always assume the high voltage (HV) battery and associated components are energized and fully charged.

Exposed electrical components, wires and HV batteries present potential HV shock hazards.

Venting/off-gassing HV battery vapors are potentially toxic and flammable.

Physical damage to the vehicle or HV battery may result in immediate or delayed release of toxic and/or flammable gasses and fire.

Do not store a severely damaged vehicle with a lithium-ion battery inside a structure or within 50 feet of any structure or vehicle.

Ensure that passenger and cargo compartment remain ventilated, i.e., open a window, door or trunk.

Notify an authorized service center or vehicle manufacturer representative as soon as possible as there may be other steps they can take to secure and discharge the HV battery.

Call 911 if you observe leaking fluids, sparks, smoke, flames or hear gurgling or bubbling from the HV battery.

Just as fires can occur during refueling a gas car, a similar risk is present when an electric vehicle is being recharged. According to a technical paper issued by NHTSA on "Lithium-ion Battery Safety Issues for Electric and Plug-in Hybrid Vehicles," the fire risk comes from several factors including electrical shorts, overcharging or over discharging the battery pack.

Other risks include exposing the battery to high temperatures or charging at cold temperatures. Excess cycling that can lead to electrochemical breakdown or defects from the manufacturing process are also factors.

While reports of electric vehicles spontaneously catching fire are rare, there have been multiple reports of fires that occur during the charging process, mostly likely involving the factors cited above. The most notable of these is the case of 13 Chevrolet Bolt fires that led to a recall of all 142,000 units produced to replace the battery packs.

A manufacturing defect was discovered as the cause, however, difficulty in detecting which battery packs were affected led to the wholesale recall. The batteries, produced by LG Chem, were also used in the Hyundai Kona EV, resulting in the recall of 4,700 vehicles.

Until Bolt owners could get the work done, GM advised them to not charge their vehicles inside overnight, limit recharging to 90 percent capacity and not let the range fall below 70 miles. GM also recommended parking the EV outside immediately after a charging session.

While the overall risk is low and the proactive nature of manufacturers to address this issue is commendable, keep in mind that from a safety perspective, a smaller battery with fewer cells and a shorter range means more safety. There have been virtually no reports of fires occurring while recharging EVs with a range of less than 200 miles. That includes vehicles like the Nissan LEAF, Volkswagen eGolf, Ford Focus Electric and Fiat 500e.

Insurance Costs

The different nature of electric vehicles also raises the question of insurance costs. What's the benefit of tax credits and incentives in lowering the purchase price of an electric car only to see those dollars disappear in a higher insurance bill?

Truth is that electric vehicles do cost more to insure on average than conventional cars. But the numbers are deceiving. Remember, the average transaction price for an electric car is $55,000, about $10,000 more than the average price of a new gas-powered vehicle.

That higher price reflects the added costs we've discussed regarding the battery pack. But also the mix of vehicles tends to skew towards higher priced luxury models. The average Tesla costs well over $60,000. Most of the new electrics coming to market tend to be higher-end models.

Since many of these vehicles cost more and use more expensive materials like aluminum in their body construction, their repair costs are going to be higher. Consequently, collision and comprehensive portions of the coverage will be higher.

Sleuthing around online supports the idea that electric vehicles cost more to insure, but the difference may not be as dramatic as overall averages suggest. The online financial site Self reports that the insurance cost in 2022 for an electric vehicle for the average American driver is $1,636, with Michigan the highest at $3,058 and North Carolina the lowest at $1,014. A gas car, according to Self, costs only $1,218 to insure, with Michigan again the costliest at $2,278 and North Carolina the cheapest at $754. According to these figures, it costs $418 more to insure an electric vehicle.

But digging down into their methodology reveals that this is more of an apple-to-oranges comparison. It starts out apples-to-apples using a 40-year-old male driver (the average American driver is 40-44 years of age) with no accidents or violations, $500 comprehensive and collision deductibles, minimum liability limits, and uninsured/underinsured coverage.

But the comparison begins to fall apart when it uses the Toyota RAV4, one of the most popular gas vehicles on the market, versus the Tesla Model 3 Standard Range. There's quite a disparity in the purchase price. A RAV4 lists for an average of just over $32,000, according to the U.S. News Cars website. The Tesla Model 3's price, according to iSeeCars.com, is just under $47,000.

For a more real world take on the actual insurance costs, let's go back to our Hyundai Kona SEL gas model versus the Kona SEL electric. Insuraviz.com, an online site that tracks insurance costs, puts the standard annual insurance cost for the Kona gas model at $1,150. The same SEL trim level in the electric version costs $1,282 on average, or $132 more.

The bottom line is shop around. In addition to the usual multi-car discounts, some carriers may have special rates specifically for electric or hybrid vehicles. Many of these companies also have prepared guides on insuring your EV. Of course, one of the best ways to secure a low rate is to have a clean driving record, take defensive driving courses and bundle your car insurance with home coverage.

Speaking of home coverage, using a 240-volt outlet, or professionally installing a dedicated Level 2 charger shouldn't impact your homeowner policy. Mercury Insurance Vice President of Property Claims Christopher O'Rourke says, "Hard-wired charging stations are usually considered part of your dwelling and thus the station likely would be covered depending upon the type and cause of a loss. Portable chargers, on the other hand, may be considered vehicle equipment, so it's a good idea to make sure to have adequate auto insurance coverage as well."

As more affordable electric cars come to the market and insurance companies build on their actual loss experience, it's likely that the insurance cost should move closer to parity. If there are significant reductions in battery cost and list prices, those costs may at some point favor electric vehicles.

Tightwad Pro Tip:

Fewer items require service or repair on an EV, so you might want to take a pass on the costly extended warranties or routine maintenance packages offered by the dealer. However, it's still advisable to stick to the regularly scheduled maintenance intervals published in your owner's manual.

11. ALTERNATE ELECTRIC AVENUES

For those a bit leery of jumping all-in on a battery electric vehicle, there are other options. You can get a taste of this technology with a hybrid with or without a plug-in option. Or you can go whole-hog science project with a fuel cell electric vehicle (FCEV).

From the Tightwad's perspective, all three have affordable options, although fuel cells are currently California-only propositions. Advantages of a hybrid include that it capitalizes on an electric drivetrain's ability to provide instant torque for smooth, swift acceleration, and the capture of the vehicle's kinetic energy through regenerative braking to help stretch fuel economy.

While having enhanced performance and efficiency from the battery and electric motor, hybrids retain the inherent advantages of an internal combustion engine's longer range and rapid refueling infrastructure. They allow you to enjoy pure electric operation in low-speed stop-and-go driving and yet be capable of taking an extended road trip on a whim. The downside is that hybrids essentially have two drivetrains that, while complementary, also bring greater cost than just having one or the other as a power source.

Two drivetrains also mean more cost. Hybrids typically have

higher sticker prices than non-hybrids. Plus, hybrids still require routine service including oil changes and general maintenance associated with internal combustion engines. Let's take a closer look at each of these alternatives.

Hybrids

Standard hybrid vehicles typically mate a 4-cylinder engine with an electric motor and a battery pack. There is no need to plug it in since the gas engine and the regenerative braking from the electric motor work in concert to keep the battery's state of charge between 20 and 80 percent capacity.

Hybrid systems can run in series or parallel. A series hybrid is powered by an electric motor with the current supplied either by the battery pack or the gasoline engine acting as a generator. A parallel hybrid feeds output of both into the same transmission, swapping propulsion duties back and forth between gas and electric, or using both sources, as needed.

This is an important factor in how the vehicle is used. A series hybrid is more efficient, especially in stop-and-go city driving since the engine runs at more consistent speeds to supply current. A parallel hybrid tends to be more efficient at freeway cruising speeds as shown by a conventional car's higher highway vs. city EPA ratings.

In either scenario, a hybrid's virtue comes from its electric assist in lower speed operation – that's why a hybrid's city fuel economy rating is usually as good as or higher than its highway rating. That difference is evident in the EPA ratings of the 2022 Toyota Prius, which is rated at 58 mpg city/53 mpg highway for a combined rating of 56 mpg.

Speaking of which, at one time the word most closely associated with hybrid was Prius. Introduced in 2000, the Prius soon evolved into a second-generation model that became the save-the-planet vehicle of choice for Hollywood. Leonardo DiCaprio and other celebrities caused a bit of a stir arriving at the red carpet in them for the 2003 Oscars.

It's currently in its fourth generation, but it's no longer the hybrid sales king for Toyota as when it came in body styles ranging from a slightly smaller Prius c hatchback to the wagon-like Prius v. In 2012, more than 236,000 were sold, a figure that dropped to 59,000 in 2021 because of wider hybrid availability both within Toyota's lineup and from other competitors.

The Toyota Prius became synonymous with hybrid technology.

The Prius c and v are no longer offered. Even though the Prius family is smaller, the brand has hybrids in nearly all its model ranges. No manufacturer has embraced the technology as completely as Toyota has.

The Prius is all-new for 2023.

Still, the Prius remains a standard bearer and carries an affordable sticker of about $27,000. The more conventional Toyota Corolla Hybrid, which rates an EPA 53 mpg city/52 mpg highway, starts around $23,000. As with most hybrids in Toyota's lineup, they tend to run about $2,000 more than their non-hybrid counterparts.

Also in the car lineup is the Camry Hybrid, which delivers 51 mpg city/53 mpg highway for just under $30,000. You can also get a hybrid full-size sedan in the Crown starting at $39,950 and net 42 mpg city/41 mpg highway. In its SUV lineup, the three-row Highlander Hybrid comes in at $40,000, about a $4,000 premium over the standard model. It's good for 36 mpg city/35 highway.

 The compact RAV4 lists for about $30,000 and it gets 41 mpg city/38 mpg highway in front-drive. Two other models offered by Toyota, the Venza two-row midsize SUV and the Sienna minivan, are hybrid only. The former costs under $34,000 and rates 40 mpg city/37 highway, while the Sienna gets 36 mpg both city and highway and starts at just over $36,000.

But that's not to say you can't find dedicated hybrid technology from other makes at Tightwad prices. Hyundai offers four hybrids, two of which start at less than $30,000.

On the car side is the Elantra Hybrid Blue is about $24,000 with an EPA rating of 53 mpg city/56 mpg highway. The midsize Sonata has a starting base price of about $28,000 and delivers 50 mpg city and 54 mpg highway.

The other two hybrids from the Korean automaker are SUVs: the $31,000 Tucson and the $36,000 Santa Fe. Both come with all-wheel drive. The Tucson garners 38 mpg on both city and highway cycles, while the Santa Fe is good for 36 mpg city and 31 mpg in highway driving.

Honda has two hybrids that are part of the broader Accord and CR-V ranges. The Accord Hybrid, which starts at $27,720, carries an EPA rating of 48 mpg city and 47 mpg highway. The CR-V compact SUV, which comes in either front- or all-wheel drive, lists for nearly $33,000 and rates 43 mpg city/36 mpg highway.

Kia has three standard hybrids on the market, the Niro, Sportage and Sorento. The Kia Niro Hybrid costs $26,590 and delivers 53 mpg city and 54 mpg highway. The Sportage costs about $27,500 and nets 42 mpg city/44 mpg highway. The larger Sorento SUV which starts at about $37,000 is EPA rated at 39 mpg city/35 mpg highway.

Ford also offers two hybrid models, the all-new Ford Maverick compact pickup truck and the Escape compact SUV. The base front-drive Maverick starts at just over $22,000 and its hybrid powertrain is rated at 42 mpg city and 32 mpg on the highway. The Escape Hybrid is $16k more with an EPA rating of 44 mpg city and 37 mpg highway.

The base model 2022 Ford Maverick is a hybrid.

The obvious benefit of a standard hybrid car or truck is its higher fuel economy. You can find them in pricier models, particularly among luxury brands, and more recently in mid- and full-size trucks and SUVs. Often these vehicles use "mild" hybrid technology that integrates an electric motor into the vehicle's transmission supported by a small battery pack. This approach is designed to boost performance of smaller V6 engines to replace larger gas guzzling V8s. The fuel economy gains, though significant, aren't as dramatic as in smaller vehicles that rely more heavily on separate electric motors and larger battery arrays.

Hybrids are a good solution for Tightwads looking to get more mileage from a gallon of gas that is getting more expensive by the day. These vehicles tend to get lost in all the hoopla surrounding not just electric vehicles but their plug-in hybrid siblings. As a result, they typically are not in the same demand, meaning there may be a deal or two to be had here.

Because there's no need to plug them in, hybrids hold appeal to those who live in apartments or condos with limited access to an outlet. Also, since they refuel like regular cars, range is not an issue when it comes to road trips. But because these vehicles can't be plugged in or operated in pure EV mode for an extended period, they are not eligible for the federal electric vehicle tax credit or other similar local incentives.

Plug-in Hybrids

Also known by their PHEV acronym, plug-in hybrids combine the best of both EVs and internal combustion engine vehicles. A PHEV not only provides more than a taste of what it's like to drive a pure electric, it also offers the convenience of traditional car ownership. It's the perfect staff car for uncertain times, especially during the lengthy transition the industry is making from gasoline to electric power.

Two cases point out the advantages of a PHEV over hybrids and pure electrics. When the Colonial Pipeline was shut down in 2021 by hackers, some Mid-Atlantic and Southern states were hit with fuel shortages. It wouldn't be a problem for a plug-in hybrid owner to recharge overnight and have some limited range at a time when people were lining up to get gas.

Meanwhile, across the country in California, PHEVs are proving their worth twice over. First, the better gas mileage and pure electric range serve as a hedge against skyrocketing gas prices. But also, during blackouts that can leave electric car owners stranded, PHEVs can still get around.

While PHEVs utilize the same basic technology as a standard hybrid, the battery packs are larger but not quite as big as on a full electric. These vehicles can be plugged in at either Level 1 or Level 2 to replenish the charge. Generally, a 120-volt Level 1 session will take six hours or so, while Level 2 can get the job done in half the time. Once fully charged, PHEVs can travel from just under 20 to over 50 miles in pure electric mode (depending on the model) before the hybrid powertrain kicks in.

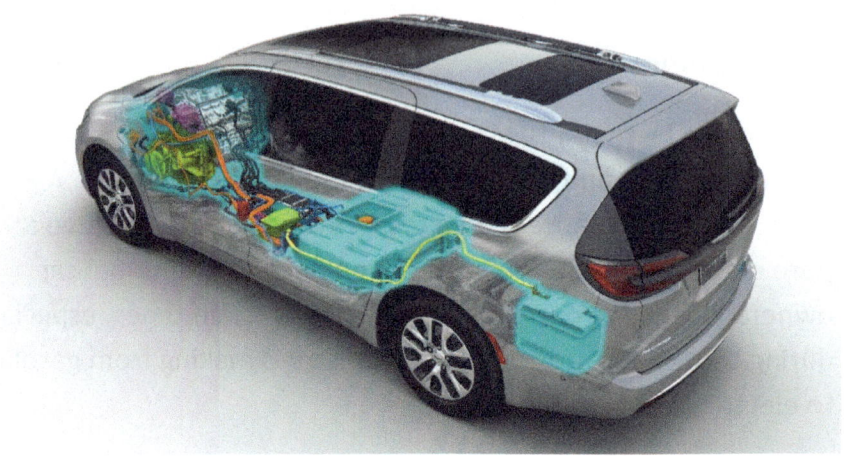

The Chrysler Pacifica PHEV features V6 power.

Because these vehicles function as electrics part of the time, they are eligible for the federal electric vehicle tax credits on a sliding scale determined by the size of the battery. In fact, some PHEVs, like the Chrysler Pacifica Hybrid, get the full $7,500 credit because of that calculation.

The Pacifica is the only PHEV of its type (remember, the Sienna comes as a conventional hybrid only). It starts at $46,978, nearly $10,000 more than the non-hybrid model. It travels up 32 miles on a charge.

As a result of their plug-in capability, larger battery capacity and more powerful electric motors, PHEVs cost considerably more than both their hybrid and non-hybrid counterparts. The tax credits, local incentives and perks like carpool lane access can tip the affordability scale in their direction.

Toyota is a good example of the range and prices to expect when shopping for a plug-in. The Prius Prime starts at about $29,000 and can travel up to 29 miles on a charge. Unfortunately, it's no longer eligible for federal tax credits, but local incentives still apply. This is bad news for RAV4 Prime buyers who are no longer eligible for the full $7,500 credit. While it costs 10,000 more than a RAV4 Hybrid and $14,000 over the base non-hybrid model, it can go a remarkable 42 miles between charges.

Hyundai offers two plug-in hybrid options in the Tucson and Santa Fe. The Tucson PHEV retails for $35,400, goes 33 miles in pure electric mode. The larger Santa Fe plug-in has a 30-mile electric range, starts at $40,000. Neither one is eligible for tax credits because of recent legislation.

The 2023 Kia Sportage offers a plug-in hybrid option.

Kia currently has three plug-in models on the market, Niro, Sportage and Sorento. The compact Niro costs about $34,000, more than $7,000 more than the standard hybrid model. It can go

33 miles in pure electric mode. The compact Sportage stickers for under $39,000 and has an electric range of 34 miles. It costs about $10,000 more than the standard Sportage hybrid.

The three-row Sorento retails for nearly $50,000 and has an electric range of 32 miles. It will set you back $14,000 over the standard Sorento hybrid. Again, all three are no longer eligible for federal tax credits, although local incentives may still apply.

Ford employs plug-in hybrid technology on its Explorer and Lincoln Corsair and Aviator, but all those models cost more than $50,000. A more affordable PHEV is the Escape Hybrid which starts at $38,500, goes 37 miles on a charge and is eligible for a $6,843 tax credit.

The Outlander PHEV has room for up to seven passengers.

Mitsubishi continues to offer a plug-in hybrid version of its compact Outlander SUV. The PHEV carries an MSRP of $40,000 and travels 38 miles on a charge. This three-row SUV is no longer eligible for the $7,500 federal tax credit.

Subaru offers a plug-in version of its Crosstrek. Starting at about $35,700, this subcompact hatchback, which comes with standard all-wheel drive, can travel 17 miles on a charge. It's no longer eligible for the federal tax credit and costs $7,000 more than the non-hybrid Limited model on which it is based.

The MINI Cooper Country ALL4 plug-in hybrid retails for $41,500. This all-wheel drive small SUV travels 17 miles in pure electric mode. It no longer qualifies for federal tax credits, although local incentives apply. The PHEV represents a $4,000 premium over the standard Countryman ALL4 model.

There are plenty of plug-in hybrid options above the $50,000 price point, most notably among luxury makes including BMW, Bentley, Mercedes-Benz, Porsche, Audi, Jaguar, Volvo and super exotics including Ferrari. Many of these are SUVs, which also include off-roaders like Land Rover or Jeep with its Wrangler and Grand Cherokee 4xe PHEV models.

When shopping for a used PHEV, the Chevrolet Volt is a good

The Chevy Volt can go more than 50 miles on a charge.

place to start. Two generations of this sharp looking small sedan were built from 2010 to 2019. Second generation models dating from 2016 boasted up to 53 miles of pure electric range.

Rarer, but no less impressive, is the Honda Clarity PHEV which was sold in a limited number of states before going out of production in 2021. This sister car to the Clarity Electric and Fuel Cell models could travel 48 miles in pure electric mode while carrying five passengers in mid-size comfort.

Ford sold a Fusion PHEV under the Energi badge.

Before Ford abandoned the mid-size car market, it built a plug-in version of its Fusion called the Energi. Last sold in 2020, the Fusion plug-in hybrid was good for an all-electric range of 26 miles and boasted a total range of 610 miles on a full tank of gas.

Ford also built a people-mover called the C-Max which looked like a cross between a station wagon and minivan. While it was tall like a van, it retained conventional swing-out doors. Sold through 2017, the five-passenger C-Max could go 19 miles on a charge.

The Ioniq PHEV was offered from 2018-2022.

Prior to the launch of the all-electric Ioniq 5, Hyundai offered a family of electrified compact cars also using the Ioniq name. In addition to a full-electric version, there as the Ioniq Blue conventional hybrid and the Ioniq Plug-in Hybrid. This model can go 29 miles between charges and has a full gas/electric range of 620 miles.

Fuel Cell Electric Vehicles

While a plug-in hybrid is a good gateway vehicle to the EV ownership experience, fuel cells are way over at the other end of the spectrum as a leading-edge technology that transcends battery power. Fuel cells essentially strip hydrogen of its protons and electrons to generate electricity.

A fuel cell contains an anode and cathode separated by membrane containing an electrolyte solution. Hydrogen enters on the negative anode, passes through the membrane where its protons and electrons are separated. The positive electrons travel up through the positive cathode as current. The remaining hydrogen atoms combine with the oxygen used to strip the electrons from the hydrogen gas as water. The only other byproduct from the process is heat.

Each cell is analogous to a battery cell, and they are stacked in arrays to produce the energy needed to drive an electric motor. Unlike a battery that needs to be recharged, a fuel cell will produce current as long as it is fed hydrogen from on-board fuel tanks, meaning it can be refueled relatively quickly along the lines of a conventional gas car.

In fact, some early fuel cell prototypes were equipped with reformers that would convert either gasoline or methanol into hydrogen that would in turn produce electric energy in the fuel cell stack. That technology, designed to take advantage of the current gasoline refueling infrastructure, was abandoned as too costly. It was like equipping each vehicle with its own refinery to make gasoline from crude oil.

So where does hydrogen come from? It is a byproduct of the refining process, so oil companies can produce it. It can also be reformed from natural gas and, through electrolysis, be produced by splitting water into its hydrogen and oxygen components. While it's a clean power source, it's also not particularly dense in its natural state, so it needs either to be liquified at extremely low temperatures (which makes refueling a challenge) or compressed under very high pressure (10,000 psi) as a gas, which is currently being used for the three fuel cell vehicles on the road in California. Refueling takes three to five minutes.

As a new technology, it's expensive. But that doesn't mean there isn't an opportunity for the Tightwad to get a deal. If you live in California in an area close to that state's hydrogen filling stations, you can choose from either a Toyota Mirai or Hyundai Nexo SUV.

Toyota Mirai is a fuel cell powered family-sized sedan.

The 2022 Toyota Mirai is a family sedan sized between the Camry and Avalon/Crown. Toyota currently offers it in two trim levels to Californians, an XLE priced from $49,500 and a loaded Limited model at $66,000. The better deal is the XLE which, with a California rebate of $5,000, brings the effective purchase price down further.

The XLE, thanks to its fewer features and lighter weight, can go 402 miles on a tank compared to the 357 miles of the more fully loaded Limited and delivers over 70 miles per kg. Throw in $15,000 of hydrogen refueling over the first three years of ownership and you have a sweet deal.

The only fuel cell SUV on the market in California is the Hyundai Nexo. The Nexo, which can travel 380 miles on a tank of fuel, starts at $59,435. But Hyundai is offering up to $25,000 in bonus cash, reducing the price to just under $35,000. That, plus state

The Hyundai Nexo offers fuel cell technology in a compact SUV body.

incentives, effectively lowers the price of in-stock Nexo models to less than $30,000. Like the Clarity and Mirai, Nexo comes with $15,000 in free fuel for three years, and both are eligible for HOV carpool lane stickers.

Tightwad Pro Tip:

There are some discontinued PHEVs that might be worth a look on the second-hand market. While these vehicles are not eligible for tax credits or incentives, their ability to travel in pure EV mode and deliver good fuel economy are plusses.

12. WHAT THE FUTURE HOLDS

If the 21st century hasn't brought flying cars, then we'll just have to settle for electric vehicles instead. The current state of the art is creating a new generation of quiet, powerful and fun-to-drive vehicles with usable range. Electrics come in all sorts of body styles, from humble compact hatchbacks to large SUVs with tremendous off-road capability. Prices range from about $30,000 to well north of $100,000.

The industry is well along the road to mainstreaming the technology. More than just a means of getting from Point A to Point B, electric cars are packed with technology that informs, entertains and protects its occupants with a wide range of features that include navigation, digital content and driver assists. Much of this new technology is shared with conventional cars. But there are other innovations coming to the electric space.

Just as wireless phone charging has become a thing, inductive charging of electric cars may eliminate the need to plug in. You simply park your car over an inductive pad and walk away as your battery is recharged as if by magic.

More than just making EVs more convenient, inductive charging technology just might be a lifesaver for the absent-minded who forget to plug in overnight.

BMW's concept for wireless recharging uses an inductive pad that the driver simply parks over in order to initiate a charge session.

Electric cars are well on the way to not only drawing electricity to recharge the batteries but also discharging that power to electrify your home or powering up a drained battery in another EV.

The concept of electric vehicles as a mobile power source has been introduced on the Ford F-150 hybrid pickup truck and is an integral feature in the all-electric F-150 Lightning.

This functionality, also known as V2L or Vehicle to Load, converts the battery's direct current to alternating current that can be used to power electronic devices, appliances and power tools as well as recharge e-bikes and scooters.

Two electric vehicles on the market now, the Kia EV6 and Hyundai Ioniq 5, offer V2L outlets located in the rear seating area and alongside the charging ports. The ability to act as a mobile power source adds another layer of utility to an electric vehicle beyond simple transportation.

Electrification is also taking hold in off-road and recreational vehicles. The GMC Hummer and Rivian R1T and R1S pickup and SUV stress that their capabilities extend far beyond where the

Vehicle-to-vehicle charging is the latest innovation in EV technology.

road ends. Recreational vehicle makers like Airstream have revealed concept motorhomes and campers that showcase EV technology. In addition to serving as a home away from home, RVs are uniquely positioned to play a larger role in managing the use of electricity.

An extension of V2L is V2G, vehicle to grid, where an electric vehicle's battery pack becomes a means to capture excess energy produced during off-peak hours. That stored energy is later tapped as a source of supplemental power during peak periods to level out generating loads.

While much of the V2G discussion centers around electric cars, imagine a camper or RV capable of being used as a home battery system like Tesla's Powerwall. Its purpose is to harvest electricity from solar panels during the day or store lower-price off-peak grid electricity for later household use.

Even better is that the technology makes recreational vehicles, which often sit idly awaiting the next trip, useful during this

Electric vehicles can help balance energy loads by being able to store solar power for use at night.

downtime. When it's time for its intended use, an electric RV or camper may be ideally suited for leisure travel. Trip segments within the vehicle's range end at campsites with electrical hookups for overnight recharging. And while electric vehicles are generally not suited for towing, a camper equipped with a battery pack and a small electric motor would ease the load on the towing EV while providing additional range.

Affordability Issues Remain

As tantalizing as the benefits are of electric vehicles, from their smooth, effortless operation to their reduced maintenance needs, affordability remains an issue. The mass market for automobiles in America wasn't created by the existence of luxury brands like Cadillac, it was built on the back of affordable cars like the Ford Model T. As we have seen, there are affordable alternatives on the market, but their numbers and unsubsidized costs still put them at a disadvantage when compared with the list price of traditional automobiles.

Until electric vehicles are accessible to a wider range of new car buyers, a wholesale move away from internal combustion cars may never happen.

At the heart of the affordability question is battery technology. Strides are being made in bringing down the cost, but challenges remain. Current lithium-ion battery technology uses many minerals that are expensive or rare. Because of these high costs, little is to be gained from scaling up production – the run-up in value of these materials due to increased demand may only serve to push their commodity prices even higher. Also adding to uncertainty are supply chain and labor issues involving China and more recently, Russia.

According to the Institute for Energy Research, "In 2019, Chinese chemical companies accounted for 80 percent of the world's total output of raw materials for advanced batteries. China controls the processing of pretty much all the critical minerals – rare earth, lithium, cobalt and graphite. Of the 136 lithium-ion battery plants in the pipeline to 2029, 101 are based in China. The largest manufacturer of electric vehicle batteries with a 27.9 percent market share is China's Contemporary Amperex Technology Co. Ltd. (CATL) founded in 2011."

Lithium-ion batteries require cobalt, nickel, graphite, and manganese for production. The IER says China's share of graphite reserves in 2019 was 24 percent, and 64 percent of the world's processed supply came from that country.

While two-thirds of the global supply of cobalt comes from the Democratic Republic of Congo, IER says China also controls 14 of the largest mines there. And over 80 percent of companies that refine cobalt are also Chinese.

IER adds that "China mines only six percent of the world's manganese but refined 93 percent of it in 2019. Most manganese supply is concentrated in South Africa, followed by Australia and Gabon. North America produced zero manganese. Ukraine has a small operation, but it is not capable of producing feedstock for the battery supply chain."

Unlike the other minerals, the nickel mining industry is spread around the world with 35 percent of the chemical processing outside of China. Electric vehicles account for about 7 percent of overall nickel consumption today, but Bloomberg New Energy Finance Ltd. predicts demand for the metal will triple for battery use to 1.5 million tons per year by 2030. Putting additional pressure on nickel prices is the fact that Russia accounts for about 17 percent of the world's supply.

As a result, the current state of the art in battery technology will not get us where we need to be in terms of having enough affordable and sustainably sourced materials required for lithium-ion battery production.

Solid State Breakthrough?

An often talked about yet unrealized breakthrough in battery technology needs to happen. It may come by substituting plentiful and cheaper materials in current lithium chemistry. Experimental approaches have tried the use of ammonia, sodium, magnesium, and fluoride alternatives, but none has yet to pan out.

The answer may be solid-state technology that uses fewer rare minerals and metals. Rather than using liquid or gel electrolytes that also heat up during charging and discharging, solid state technology promises to take a charge quicker and generate less heat in the process.

This approach also would produce batteries that have a longer life and better range than the ones available today. The question remains at what cost and how soon.

In a technology outlook report titled "Are Solid State Batteries the Holy Grail for 2030?" DNW Consultants concludes that during the initial development phase, "solid-state technology is estimated to have high cost varying in the rage of [around] $800 to [around] $400 per kWh by the year 2026."

Currently, lithium-ion technology is about $137 per kWh. The company adds, "The comparatively high cost may significantly hinder production and uptake of solid-state batteries."

But, the report says, "With improved power density and lower cost, our Energy Transition Outlook forecasts that 50 percent of all new passenger vehicle sales in 2032 will be electric. With a market breakthrough of solid-state batteries, it's likely these numbers will grow even further."

Toyota is planning to launch its first solid state battery in a car by 2025, but it will be in a Lexus, hardly the beginning of a mass-market revolution on the scale of a Model T. Other makers investing in solid state technology are Volkswagen, BMW, Ford, General Motors and Hyundai.

Think Small

Still, by all measures, we won't be seeing this breakthrough in the immediate future. So, we're left with current lithium-ion technology. And if manufacturers are truly serious about widening the appeal of electric vehicles, the market could use more Tightwad-friendly EVs like LEAF and Kona.

The key to affordability, whether we have solid-state batteries or not, is to get over the relentless push to increase range to 600 miles or more on a charge. Unlike a traditional car, where you can make a tank bigger at minimal cost, increasing range from bigger batteries comes at a huge cost in electric vehicles.

Of course, work will continue to improve current technology to make battery packs that are more compact and lighter while delivering greater range. But the expectations should be for incremental steps rather than quantum leaps. It's a question of adjusting expectations to align with affordability.

Many of the makers committed to a major shift towards electric vehicles are luxury brands. As we've so far seen, volume manufacturers introducing electrics seem to be skewing their efforts towards a more well-heeled clientele. Given the higher costs of current electric vehicle technology, it's understandable.

Still, recent launches of pricey electric cars, trucks and SUVs are echoes of the large, flashy cars with big fins and massive chrome bumpers that marked the end of the 1950s. It wasn't until an imported air-cooled economy car called the Volkswagen Beetle with its ad tagline suggesting we "Think Small" that affordability took center stage. Perhaps manufacturers need to take a step back and think small to provide affordable electric alternatives for average car buyers.

Thinking small in the EV space may mean going only 200 miles on a charge, rather than having to pay a hefty premium to go twice as far. If there are improvements both in the availability of fast charging, the sacrifice in range won't matter much if it takes less than 10 minutes to get 80 percent of that distance back. Greater convenience is the key.

Battery electrics have already proved that they can meet the needs of many in a wide range of uses in both urban and suburban settings. Manufacturers like Nissan, Hyundai, Kia and Chevrolet have shown that they can build and sell electric vehicles for less than $40,000, with more on the way. Chevrolet promises an electric version of its Equinox priced at $30,000. Recently, GM announced a tie-up with Honda where the auto giant will supply its Ultium battery technology to the Japanese make. Honda, in turn, would build the smaller and more affordable electric cars and SUVs for both companies to sell starting as early as 2027.

Electric cars are not a transportation panacea. But they offer another way to meet everyday needs for personal mobility. While electric cars will take a greater share of overall sales, they will not totally replace internal combustion vehicles in the foreseeable future. An argument can be made for both technologies. Understanding the differences and how each can best serve your particular needs are key to making the right purchase decision.

Make no mistake, electrification of the automobile is here. The success of EVs, however, will not be because of some lofty pronouncement of an all-electric future by a manufacturer or politician, but rather by the decisions made by car buyers in the market. In the end, the true measure of success is that there are enough electrics cars on the market priced right to appeal to the Tightwad in all of us.

ACKNOWLEDGMENTS

This book wouldn't be possible without the love, support and efforts of my partner for life, Jane. Collaborator and keen-eyed editor, she's been with me every step of the way. Many thanks to photographer Guy Spangenberg for many of the photos and to cartoonist Henry Payne for adding a touch of whimsy. I'd also like to recognize my many colleagues for their advice including Jack Nerad, Richard Baron, Tim Gallagher, Ralph Hermans, Jim Finn, Mike O'Brien and Roger Conner. I'm grateful for the love and support of our children, Amy, her husband Joey, Stephen and his significant other, Emily. Also, thanks to friends for their support including Robert, Eric, Daryll, Ray, Don, Lyndon, Alex, Scott, Chip, Gary and Tim. Finally, all thanks to God for the inspiration, grace, and gifts to create this work.

ABOUT THE AUTHOR

Matt DeLorenzo has made his career writing about cars. A former editor at *Autoweek* and *Road & Track*, he's written many books. Among them are *Legendary American Cars, Modern Chrysler Concept Cars, Corvette Dynasty, Dodge 100 Years* and books about the Volkswagen New Beetle, PT Cruiser, Hummer H2 and 200 Ford Mustang. He lives with his wife, Jane, in Orange County, California.

Made in the USA
Coppell, TX
31 May 2023